"This book is well-written, accessible to readers without an academic background, and entertaining. It's a quick yet informative read."

— Nalini Elisa Ramlakhan, Philosophy PhD student at York University

"More succinct and accessible than Piketty."

— Leo Eutsler, retired carpenter

"These are pressing social issues indeed. I found the dialogue method engaging and good leavening to the statistical information. We can do better as a society!"

— Phil M. Lew, builder and designer

"Many times I've struggled with your academic doorstoppers. Now you write something readable—and you're telling me that publishers don't want it?!"

— Grant M. Morrison, author's cousin

ERIC W SAGER

THE
PROFESSOR
AND THE
PLUMBER

Conversations about Equality and Inequality

ILLUSTRATIONS BY HANNA MELIN

 FriesenPress

Suite 300 - 990 Fort St
Victoria, BC, V8V 3K2
Canada

www.friesenpress.com

Copyright © 2021 by Eric W Sager
First Edition — 2021

All rights reserved.

No part of this publication may be reproduced in any form, or by any means, electronic or mechanical, including photocopying, recording, or any information browsing, storage, or retrieval system, without permission in writing from FriesenPress.

ISBN
978-1-03-910556-0 (Hardcover)
978-1-03-910555-3 (Paperback)
978-1-03-910557-7 (eBook)

1. Inequality, Social Justice

Distributed to the trade by The Ingram Book Company

TABLE OF CONTENTS

1. LET'S TALK	1
2. EMILY, A WORKER	5
3. MICHAEL, A CEO	10
4. THE RISE OF INEQUALITY	15
5. DOES IT MATTER?	25
6. DESPAIR AND DEATH	32
7. WE NEED INCENTIVES	37
8. FIXING THE PROBLEM	43
9. WHAT WE VALUE	51
10. EQUAL OPPORTUNITY	57
11. REWARDS AND MERIT	62
12. PAY EQUITY	68
13. WHAT IS PROPERTY?	71
14. WHAT IS FAIR?	77
15. WELFARE STATES	83
16. EQUALITY AND CAPABILITY	89
17. EQUALITY AND COMMUNITY	95
18. SUCKING UP WEALTH	101
19. RISING ABOVE MYTHS	105
20. ALTRUISM AND A GAME	112
21. AN EQUALITY MANIFESTO	118
SOURCES	126
THANKS AND FINAL THOUGHTS	151
ABOUT THE AUTHOR	155

1. LET'S TALK

"YOU WANT A beer?" asked the professor.

"Maybe not. Had a few yesterday with the guys after work," said the plumber.

"Well, let's eat. I've made lasagna. And if you change your mind about the beer, I'll limit you to one."

"Whatever you say, boss. Umm, do you mind . . . can we turn down that flute music? It's not helping."

"Huh! You *are* in a bad way, aren't you? So can we talk while we eat?"

"Yeah—you talk while I listen and take another Tylenol. You're always jawing about something."

"Guilty as charged! So I think it's my turn to suggest a topic, isn't it? And this one might take a while. Unless you can talk all night like the ancient Greeks did."

"No way! Forget about the Greeks. So what are we talking about?"

"I want to talk about *inequality*."

"Inequality? You mean economic inequality? Okay with me. But oh my God! That's a monster of a topic. My head's hurting already. . . . And hang on—before we get into it—can you tell me *why* you want to talk about inequality? What's the point?"

"Well, I've been thinking about inequality for a while, you know. And I have to teach the subject. So I can try out my ideas on you, can't I? But let me ask you a question first. You follow the news, right?"

"Umm . . . well, not every day. It's too depressing! Everything is so screwed up. . . . Crazy politics, climate change, and a pandemic that won't go away . . . I'd be better off watching horror movies."

"So there's no way out of this mess?"

"Maybe. Like, in your dreams."

"No, little buddy. Not in my dreams. You want a way out of the confusion? If so, we've got to think—before we read the news and get depressed. We need some principles—some rock-solid beliefs. Then we might be able to make some sense of it all and do something about it. And besides, inequality is really important. It's about the kind of world we want to live in."

"Okay, we can talk about inequality. But I need to get what you're trying to say. So don't go all academic on me."

"No way! You know me better than that. We know how to talk to each other."

"Yeah but sometimes it sounds as though you're just talking to yourself."

"Ow! That hurts. Give me a chance, will you? Besides, you love ideas, and you like arguments. You're the most argumentative person I know."

"Okay . . . lead us out of confusion. If you can."

"So let me start with a bit of a professor speech. Do you know what a Gini coefficient is? It's spelled with a 'g'— g . . . i . . . n . . . i . . . but it sounds like your girlfriend's name, Jeannie."

"No I don't know what it is. Should I?"

"It's where we start. The Gini coefficient is the most common measure of inequality. It's a way of measuring and comparing the distribution of incomes."

"I hate statistics. It's just numbers. Math was the only course I failed in high school."

"Well, you just gave me a statistic."

"No, I didn't."

"Yes, you did. You stated a relation between two amounts: how many courses you took and how many you failed."

"And you are a smartass university professor."

"Anybody who doesn't know us would think you were being rude."

"Yeah, but you know me. That was me paying you a compliment. And I don't throw bouquets very often."

Fair distribution: a problem for plumbers and professors

"So let me explain! You failed one course out of N courses, where N is the total number of courses you took in high school. Let's say you took 25 courses throughout high school, so N is 25. The ratio is 1 over 25 and that equals 0.04. The ratio is really small, so I'm convinced that your failure in math was really exceptional."

"I just zoned out during class."

"So you didn't actually know if you were good at math, did you? But you still use numbers. I bet you use numbers at your work all the time. And you did math courses later to catch up, didn't you?"

"I don't need any Gini coefficients to know what inequality is. It's where you've got lots of stuff and I've got hardly anything. . . . Though maybe it depends on what stuff we're talking about."

"Yeah, I think you're right. . . . We'll have to talk about that later. . . . And inequality is not just a difference between you and me, is it? If inequality exists, it's a difference between large numbers of people. Some have more and some have less desirable goods."

"Yeah, like plumbing. I see it all the time. Go to a low-cost rental building. Leaks, corroding pipes, holes in the walls—and landlords won't replace the Poly-B pipes. That's inequality for you."

"So you know something about inequality, don't you? You've seen it. And as I've told you before, I learn things from you too. You're my teacher, cousin, as much as I am yours."

"So you say."

"I do—which might be a mystery to a lot of people (except Socrates, of course)."

"But you know a hell of a lot more than I do."

"Really? What is knowledge anyway? And isn't there a difference between knowledge and wisdom?"

"Okay, prof. But don't give me numbers. I want to hear about people. Stories about real people."

"Are you okay with stories that include dollars and dollar amounts?"

"Oh all right, if you insist."

"So let's talk about two people: Emily and Michael. I made them up, but I did my research. So everything about them is based on reality."

"Can I have some more lasagna?"

2. EMILY, A WORKER

"COULD YOU LIVE in a Canadian city on $65 a day?" asked the professor.

"Only $65! That's about $2,000 a month. No, I couldn't live on that, seeing as how much I pay in rent," said the plumber.

"Emily lives on $2,000 a month. So let me tell you about her. You can read my lecture notes, if you want."

Emily is a single mother. As of January 2020—just before the pandemic hit—she lives with David, her 8-year-old son, in a town in southern Ontario. Emily and her son are a small family of two—they're one of the more than 1.6 million single-parent families in Canada.

Emily knows she's lucky to have a full-time job. She works in a warehouse, where she "picks and packs" online orders. It's hard work—you're on the move all the time—but at least it's a steady job. Emily works hard and never says a word when the supervisor gets nasty. She can't afford to lose this job. She earns $2,000 a month.

About $1,100 a month goes to rent. Emily and David live in a two-bedroom apartment in one of the older buildings in town. Hydro bills are high in the winter, but they average about $60 a month over the year. Emily's housing costs are more than half of her income.

"Yeah, and I bet she'd pay more in rent if she lived in Toronto."

"For sure. Probably more than $2,200 for a two-bedroom apartment."

Emily spends about $300 a month on food. She tries to shop at Superstore, but it's a long way from their home, so it takes time. She has a bus pass, which costs $81 a month, but she's heard that it will soon be going up to $95 a month.

It's not easy getting the food bill down to $300. That takes time and effort. It means finding coupons or using a food app to find good prices. A friend found her a cheap cell phone plan for $32 a month. Her cable TV and Internet plan started at $41 a month, which sounded like a good deal, but it has been creeping upward and is now closer to $80 a month.

David helps his mom with the shopping every week. It's hard, but he has learned not to ask for chocolate bars or potato chips. Emily knows about nutritious food. It's important to have fruits and vegetables, but they are expensive and the prices are going up all the time. She doesn't go out to restaurants, but about once a month, for a treat, she and David go to McDonald's for dinner.

Emily would never consider going to a food bank. She doesn't need to. She's proud that she can look after her son on her own, even if it means going without expensive things for herself. She has found a few shops, like the Goodwill Centre, that sell quality clothing. She doesn't want other kids to make fun of David for wearing cheap clothes or things that don't fit.

David has a serious problem of tooth decay, and the health plan doesn't cover dentistry. She asks herself: what are we going to give up so he can go to a dentist? It might cost several hundred dollars. Do I cut back on the food? Or do I cancel the cable TV? Emily plans to apply to the Healthy Smiles program for low-cost dental work and hopes that her son will qualify.

At least Emily doesn't have to pay for daycare, now that David is in school. But she does have to pay for after-school care. The normal fee is $16 a day, but she gets a subsidy, which brings the cost down to $150 a month. Sometimes the school is a real problem. Emily receives a request for money to cover a few school trips and other activities, but she has to reply that she

2. EMILY, A WORKER

can't afford it. David will miss some of the activities but not all of them (the school has a fundraising drive to help those who can't pay).

"Umm . . . does she have a credit card?"

"Yes, she does. But she tries hard not to use it too much or to pile up debt. She's got a small debt though—she had to buy a new bed for her son recently. And the interest hurts."

Emily is doing an online college business course, trying to upgrade her qualifications. She thinks that with a diploma she might have a chance at landing a better-paying job. But she has to pay a tuition fee, of course. She works on her course in the evenings, after David has gone to bed. Her working days are long. She's up at 6 o'clock every morning, preparing food, making lunches, and making sure David gets a proper breakfast.

"Hold on! Add up her expenses so far. Does her income cover the expenses?"

"Barely. Her monthly expenditures so far are around $1,900—just below her monthly income."

She would like to buy a car, but she can't afford one. So far, there's nothing left in her budget for entertainment, for emergencies, for swimming lessons for David, or for savings. Every recommended budget plan for those with low incomes says that regular savings are essential. In ten years' time, David may want to go to college or university. But how?

Emily has a few options. She can get another job, working at night. She can apply for subsidized housing for low-income people. She can borrow a bit from her parents (they have helped her already in emergencies), but they were working people, so they live on small pensions and can't help much. Meanwhile, somehow, Emily gets by. She didn't choose to live on $65 a day.

Emily might get a raise next year (her warehouse is unionized). A 3 percent raise, if she's lucky. That would be an extra $60 a month.

2. EMILY, A WORKER

"You know what? She sounds like my old friend, Carol. She's a single mum too. She really struggles, but she's tough—she won't accept any help."

Emily is not alone. She's among the bottom 10 percent of income earners in Canada. And women like her were hit really hard by the pandemic in 2020—a huge number of women lost their jobs. But perhaps things are better for Emily than for a single mother a generation ago? She has cable TV and a cell phone, after all. The answer is "no." The real disposable income (that is, taking inflation into account) of the bottom 10 percent has not changed since the 1970s. And the total share of all incomes has actually fallen.

Emily is not among the poorest in Canada. She has a job and a place to live. What about those who are homeless, trying to live on a few dollars a day, from foraging and panhandling? On any given day, some 35,000 Canadians are homeless. Over a million have no job. And the COVID pandemic has caused an increase in homelessness—and a huge increase in the number of people who suffer from food shortages.

"So that's my story about Emily. You can relate, eh? Now let me tell you about another Canadian."

3. MICHAEL, A CEO

"COULD YOU LIVE in a Canadian city on $30,000 a day?" asked the professor.

"For sure! But who's got that much?" wondered the plumber.

Let me tell you about Michael. He is a Chief Executive Officer—a CEO. He lives in Toronto with his wife, Brenda. They have two grown-up children who have their own homes. Michael and Brenda live in a nice 5,000-square-foot house, with four bedrooms, four bathrooms, and a swimming pool. They own their house, and there is no mortgage. They pay a lot for utilities, but their housing costs, as a percentage of their incomes, are close to zero.

Michael's annual remuneration—$11 million—is around the average for the top 100 CEOs in Canada.

Michael and Brenda spend about $2,000 a month on food. That includes food for regular dinner parties (mostly Michael's business friends and associates). The chef/food shopper comes twice a week. The housekeeping service means that they do not need a live-in maid.

Michael's company is in the energy sector. He has always worked in this sector, after completing his commerce degree, and thinks of himself as "self-made" (his father owned a furniture retail business). His working days are long. Out of bed by 6:00 a.m., he reads the newspaper and always chats with Brenda over breakfast. At the office by 8 o'clock, his day begins with a conference call with regional managers, followed by a meeting with executive assistants to review the day's agenda. Much of the day

3. MICHAEL, A CEO

is taken up with meetings with the division leaders and reports from various sectors: marketing, exploration, development in wind power and biofuels, oil sands, and much more. He is often at work in the evenings or at public events. He travels a lot, sometimes by Air Canada but more often in the company jet.

Work and leisure overlap and networking occurs at the Rosedale Golf Club or in the company's executive suite at the Rogers Centre. Michael drives to work and to the golf club in his BMW.

Brenda does a lot of work with the company's foundation. She's on the board of the foundation and researches funding opportunities in the company's preferred sectors: Indigenous peoples, community support, and sustainability. Brenda drives an electric car.

Have you noticed something? Their income and spending profiles are unusual—but however you add up the spending, it would not come anywhere near $916,000 a month. Of course Michael and Brenda do not have that much in disposable income. Most of it is "tied up" in shares, stock options, perquisites (paid expenses), and such things. Only about $1.4 million of his compensation is salary, and taxes take a large portion of that salary. Of course Michael and Brenda hire accountants to limit their tax liability. In Canada, the overall effective tax rate on the top 1 percent of incomes is about 31 percent. They are still left with more than $56,000 a month, and it is hard to spend that much.

This is why the top CEOs subscribe to Global Living, Robb Report, Upscale, *and other magazines. To help them figure out how to spend their money.*

Next year Michael might get a pay raise. Perhaps 10 percent or even 15 percent—that would not be unusual. Perhaps a pay increase of $1.6 million.

Such pay raises have been going on for many years. The real income (adjusting for inflation) of the top income earners in Canada has grown by almost 4 percent a year since 1980. Their share of all incomes has also grown.

11

"There's a lesson here. If you want a pay increase, try working hard, and you just might get a small increase. But if you want to ensure you receive an increase, join the rich, preferably the super-rich, because the richer you are, the bigger your pay raise. Emily and Michael live in different worlds. What do you think? Could they change places? Even for a day?"

"No way. They would both be total benchwarmers."

"Creators of imaginary worlds can help us here. Many years ago, a Canadian author named John Marlyn wrote a great novel, *Under the Ribs of Death*. It tells us about a young boy named Sandor who walked from Winnipeg's North End into a world he had never seen before—River Heights, where the super-rich lived."

It was as though he had walked into a picture in one of his childhood books, past the painted margin to a land that lay smiling under a friendly spell, where the sun always shone, and the clean-washed tint of sky and child and garden would never fade; where one could walk, but on tip-toe, and look and look but never touch, and never speak to break the enchanted hush . . .

In a daze he moved down the street. The boulevards ran wide and spacious to the very doors of the houses. And these houses were like palaces, great and stately, surrounded by their own private parks and gardens.

"Well, yeah! Fantasyland! But I bet things weren't always happy inside those palaces."

"You're right. The sun did not always shine on the world of the rich. But Sandor's dreamy wonder is telling us something. It tells us that inequality is not just economic or material. It's also cultural and psychological. The rest of the story is about what happens to Sandor as he leaves the North End and crosses into the world of the rich. You should read it!"

"Okay, maybe one day I will. But remember—you've got lots more time for reading than I do."

"They live in the same country. Each breathes the same air. They both eat breakfast every day. But they do not eat the same bread.

They do not wear the same clothes. One may hear music on the radio; the other may hear music in Roy Thomson Hall or Covent Garden. One takes holidays at home; the other stays at an estate in Tuscany. One might ride the ferry to Toronto Islands; the other enjoys a yacht on Lake Ontario. They live in different worlds—and in different countries of the mind."

4. THE RISE OF INEQUALITY

"THOSE STORIES WERE a thrill, eh? But you were always good at telling stories. Real scary ones. Ghost stories. You used to scare the hell out of me," said the plumber.

"So now I'll scare you again. There's another way to tell the story of inequality. Economists have their way of telling it," said the professor.

"Yeah, but they don't speak my language."

"Well, they give us summaries that are easy to see and understand. Look at this graph. Here you can *see* the top 1 percent over a long period of time."

"Wait a minute. What am I looking at?"

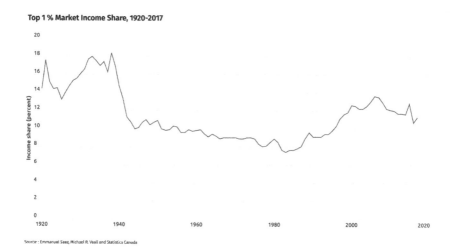

Top 1 % Market Income Share, 1920-2017

Source : Emmanuel Saez, Michael R. Veall and Statistics Canada

"The top 1 percent—one out of every hundred. So in 1939, there were around 4 million income earners in Canada. One percent of that is 40,000 people. Look at 1939. What it tells you is that those 40,000 people—only 1 percent—took home 18 percent of all the income!"

"Okay, I get that. . . . And then their share dropped. I could ski down the slope that started in 1940. What happened?"

"It's not easy to answer. It's something to do with the Great Depression of the 1930s and the Second World War. The Depression was a huge shock. All parts of the economy took a hit: capital, agriculture, trade, and labour (wages and salaries). Recovery began in the late 1930s, but then there was sudden and very rapid growth during the war. This growth was managed, and much of it was paid for by the state—the public sector—because so much was being produced for the war. There were more jobs, and wages and salaries started to recover. But capital recovered more slowly. That doesn't mean that capitalists were getting poor and workers were getting rich—far from it. It means that there was a shift in the balance of national income. Capital got a smaller share."

"Okay, but the war ended. And the top 1 percent stayed at that lower level until sometime in the 1980s. Why would that happen?"

"Now you're really getting curious, aren't you? Can I give you a few books to read?"

"Hey, cousin, I'm too blessed to be stressed. And I know you can tell me the basics."

"I can't give you a total answer. But here's a list of things that need to be considered."

- During the war, tax revenues became bigger than ever before. Tax rates on top incomes were high. After the war, tax rates declined but at first quite slowly.
- At the same time, many more workers had ways of protecting their wages, especially through trade unions and legalized collective bargaining.

4. THE RISE OF INEQUALITY

- Did the growing welfare state help? Probably not much at first, but new ways of sharing taxes across provinces may have helped.
- Other factors? Possibly the way companies controlled CEOs' salaries. And the importance of domestic markets since globalization came later. So if wages stayed high, that could help companies because more people had more money to spend on what the companies produced. Of course employers often fought against wage increases, but they could also benefit from them.

"Okay. And then inequality goes uphill, beginning in the 1980s."

"Yes, plus the rise in the top 1 percent's share was even steeper in the United States."

"I bet I know why! Competition! The world got a lot more competitive, especially with globalization. So the business world got rough. CEOs and all business leaders had to be much more brainy and better educated to survive. So they got bigger rewards. Am I right?"

"Excuse me? They got more smart? You're so young, little buddy—maybe you're too young to remember the big crash in 2008 and 2009. The myth of CEO competence got blown away.

"Let's rule out that explanation right from the start. In the postwar decades, there was plenty of competition. And nobody has ever proven that top CEOs, or other super-rich people, got smarter or worked harder in the 1990s and early 2000s than they did in the 1950s.

"Was the increase in incomes for the top 1 percent some kind of return for hard work or greater skill or greater productivity? In other words, were the rapidly growing incomes and wealth of the top 1 percent a fair reward? *No!* Economic growth was faster, productivity growth was faster, and the rate of return on investments was higher in the middle of the 20th century, when top-end incomes were relatively lower.

"After the 1980s, were the wealthy getting higher pay increases as a reward for *new* skills or higher education levels? That's hardly likely. Education levels have risen across the population, but only a few have seen their incomes rise so steeply."

"Yeah, so you say, but surely CEOs like Michael are just smarter than the rest of us."

"So what about Emily? You think that living on $65 a day doesn't take smarts? It takes a lot more than mere luck, I'll tell you that. It takes thinking, planning, making tough choices, and even research. It's damn hard work."

"Yeah, okay, but being a CEO must require a higher order of intelligence."

"But what's the point of the story about Sandor, the boy from the North End of Winnipeg? The point is that the super-rich live in a different world. You can't compare intelligence in one world with intelligence in another world. How would you do that? By what measure?"

"Well, there are measures, aren't there? Such as IQ? I bet your IQ is way higher than mine."

"That's *definitely* not true! IQ tells you almost nothing. There's no perfect correlation between IQ and levels of success. Much more important than measured intelligence are certain personality traits, inherited advantages, and cultural circumstances."

"Okay, genius, but a big CEO just has to be smart. Otherwise he wouldn't be a CEO."

"That's not a valid argument because the reasoning is circular: the cause of becoming a CEO—smartness—is also the outcome (being a CEO). You're using what you're trying to prove as the proof. So you've said nothing."

"Gee, thanks. Why is it so hard to win an argument with you? But surely CEOs have more responsibilities today. Companies are bigger. Decisions have big effects on shareholders, employees, and society as a whole. Bigger responsibilities mean bigger rewards."

"You may be right. But look at the modern corporation. Responsibility and decision-making never fall to just one person, even where management tends to be top-down. Responsibility

4. THE RISE OF INEQUALITY

is dispersed and decisions are made by teams—committees and boards—of varying sizes."

"There must be some measure of competence that helps show why the CEO succeeds and rises above the ordinary level of success in business."

"Even if there were such a measure, how could you prove that the CEOs of the 1990s and early 2000s were much more capable than CEOs of previous generations? So much more capable that this greater capability led to their huge pay increases? You can't prove that, and circular reasoning won't help. You've not yet begun to understand why only the rich get big pay increases every year, higher than the inflation rate—and why the richer the person, the bigger the increase."

"Okay, you desk jockey. I don't like it when you do all the talking, but I'll suffer a bit longer. So tell me why *has* inequality increased so much since the 1980s?"

"Are you expecting a simple answer? I hope not. Whether you like it or not, there *is* no simple answer. But let's not shy away from complexity. The world is not a simple place.

"All I can do is tell you what needs to be considered when trying to answer this question. I'll give you a list. And let's accept complexity, okay? That means you're not allowed to choose the factors that sound most convincing and say 'there's the answer!' The only way to give a complete answer would be to take each factor, justify its relative importance, and weigh it (does it have more or less influence on the rise of inequality?). Some factors will have more weight than others in any final answer. And you might have to do more than that. Ideally, you should show how each factor might interact with other factors, strengthening the combined effect of both, and then you would have to weigh the interaction!

"That's why it's very difficult to be an economic historian."

"All right, all right—but do I really need such a big, long list? You can't suddenly turn me into a historian, you know!"

"Hang in there, cousin! Aren't you curious? Besides, if you want to understand something, you need to know where it came from."

"Yeah, I suppose. Like, if there's shit in a drain, I need to know where it's coming from."

"So here's a short list of factors. Remember that recent inequality was driven by the more rapid rise of incomes among those who were already rich. The share of all incomes earned by the top 1 percent increased from between 5 and 6 percent in the early 1980s to over 10 percent between 2005 and 2008. The increase in income shares is even faster if you look at only the top 0.01 percent."

- **The Big Shift.** In Canada, and in many other countries, there has been a decline in the share of national income that goes to wages and salaries. This "labour share" fell after the early 1980s, and the share going to capital rose. Some argue that there is a direct correlation between this shift and rising inequality.

- **Bigness gets more.** Large companies have grown much larger, and the salaries of top executives grow with each firm size. There are many reasons for this, but one is the increasing ability of CEOs to control their own pay.

- **The USA effect.** Huge companies are concentrated in a few cities, and they compete with each other to hire the top executives. The Canadian economy is closely tied to the American economy. Thus Canadian CEOs compare their pay with the pay of American CEOs, rather than the lower pay of CEOs in Japan (for instance).

- **Big finance.** In many countries, there has been a growth in the importance of finance capital (bank loans, equities, bonds, derivatives). Does finance capital smooth investment in productive sectors, or does it make profits by moving money around rather than creating value? This question has led to lots of debate. But many argue that finance capital has become increasingly "rent-seeking" (seeking to increase one's wealth without creating new

4. THE RISE OF INEQUALITY

wealth). This rent-seeking fuels the rise of incomes at the top-end.

- **The decline of trade unions.** Explanations for rising inequality often emphasize the decline of conditions that existed in the "great compression" (the decades after the war when inequality was relatively low). One thing that helped maintain the labour share of income was the relative strength of unions in the third quarter of the 20th century. In Canada, union density (the proportion of the labour force that was unionized) fell from 37.6 percent in 1981 to 28.8 percent in 2014. Union density is even lower in the private sector. In European countries where union density is much higher, income inequality is lower than it is in Canada.

- **The decline of the welfare state.** There have been steep cuts to social assistance in Canada, especially in the 1990s. Also, unemployment insurance (renamed Employment Insurance) now gets paid to a much smaller percent of the jobless than it did before the 1990s. The decline in Canada's social security system has certainly made life tougher for many at the bottom end of the income scale. But how far has that decline contributed to rising inequality? This question is hard to answer. When it was created, the social security system was not intended to reduce inequality. Its purpose was to offer modest and temporary support to severely disadvantaged people.

- **Lower taxes**. Corporations and right-wing think tanks whine about taxes, and they always will. But compared to other OECD countries, Canada's tax rates on top incomes are relatively low. And they have declined. Marginal tax rates on the top 0.01 percent were over 80 percent in the early 1940s; they fell to under 50 percent by 2000, and they are even lower today. Tax evasion and tax avoidance are huge problems. But how can this affect inequality, which looks at before-tax incomes? The answer is complicated. But it is often argued that the lower tax base weakens our transfer and social security systems, and it weakens our ability to alleviate poverty and inequality.

- **Technology.** The argument is that the growth of the high-tech sector has led to greater inequality because a minority of workers with rare skills command a big wage premium. Many well-paid jobs in manufacturing, for instance, get replaced, and more workers shift into lower-paid sales and service jobs. This argument feeds into a related one about the fear of technological unemployment (workers being replaced by robots and artificial intelligence)—a subject that sparks much debate.

- **Globalization.** There is more than one argument here. One is that globalization—increasing international trade and flows of capital and knowledge—encourages greater inequality *within* developing countries. The other is that more manufactured goods come from low-wage countries, weakening high-wage manufacturing in rich countries. At the same time, rich countries export more high-tech goods, giving a boost to high-tech and high-wage jobs.

"That's a lot of reasons! And it all sounds sort of... well, mechanical. Huge crapware beyond anybody's control. Are there any people in this story?"

"You've got a point, I think."

"Yeah, in my business it's all about the gear—the pipes and cutters and machines. But if something goes wrong, there's usually some human screw-up somewhere."

The super-rich live in a different world

"So perhaps we need to talk about how people think and how they act. We need to talk about ideas and culture. In North America particularly, a 'free market' cult became dominant in the late 20th century. The idea was that markets are efficient, and we must always accept what markets tell us. Such ideas were part of *neoliberalism*. This doctrine held that market forces determined all outcomes, and any attempt to deflect market forces was doomed to fail. Many books have been written about why neoliberalism, a minority doctrine in the mid-20th century, triumphed to the extent that it did.

"Have you heard of something called *homo economicus*? It's an idea about humans. Human beings are rational and self-interested actors who seek to maximize personal utility. It followed from such ideas that inequality—even extreme inequality—was the product of market forces and nothing could be done about it. Neoliberalism had lots of influence, even on governments. The catastrophic deregulation of the financial sector in the United States is merely one example."

"So you know what? Now your list of reasons is even longer! Sounds to me as if inequality was inevitable. Unless . . . umm . . . are there any places where inequality did not increase?"

"You're right! Inequality is not the same everywhere. Which proves that it is not inevitable. Even where globalization and technological change have occurred, inequality is often lower than in Canada. In many European countries, inequality is much lower. In France, Sweden, and Japan, the share of the top 1 percent changed only slightly after the *great compression*. The lines in the graphs for those countries remained flat!"

"So, you know-it-all Professor, if you're right, extreme inequality is not inevitable. It's a choice."

5. DOES IT MATTER?

"ICE CREAM!" EXCLAIMED the plumber.

"Chocolate," replied the professor.

"My favourite. And I suppose you're gonna tell me I can't have a beer as well."

"One treat at a time, you greedy glutton."

"Next you'll be reminding me to eat my vegetables. Umm . . . remember the waterslides?"

"We had just enough to pay the bus fare, to pay to get in, and to buy an ice cream cone."

"Yesterday, all my troubles seemed so far away."

"And I believe in yesterday. . . . Our secret code."

"Hey, wait a minute, you sneaky academic. Look! You got more ice cream than me."

"Inequality again. And it's unfair, right?"

"Yeah, it sucks. So next time I get more. . . . But that's the way the world works, right? There's always been inequality. What about during the Middle Ages? The lords lived in a different world from peasants."

"True. But is there a difference today?"

"I don't see why not. Today the super-rich live in a different world from the poor."

"Ask yourself this: could the peasants see into the world of the lord? Could peasants walk into the lord's castle? Could they imagine living in that castle?"

"Probably not. But it's the same today. Ordinary people can't walk into the mansions of rich people."

"Can they *see* into those mansions? Can they see the luxury cars that the rich drive and the yachts they cruise around in?"

"Sure . . . from a distance. On television and on their computer screens."

"And what do people feel when they see those things? When they see them all the time? In advertisements that encourage people to believe that they should want them."

"Umm . . . I don't know. Maybe it will make them work harder so that they might get some of those things."

"What are the chances that they'll succeed?"

"Well, it's possible for somebody from an ordinary family to become wealthy, right? I've heard of big business types who came from poor families."

"Yes, there are such people. But that doesn't answer the question: *what are the chances?* You've got to look at it from the other direction. Take all people who are born into lower-income families—or even into the bottom 40 percent. How many will eventually join the top 1 percent? The percentage is *tiny*. Just because something is possible doesn't mean it's likely."

"Okay, but people can still hope, can't they? And hope encourages them to get educated and work harder."

"So tell me: how do people feel when their hopes are dashed? When they work really hard and get small rewards? Yet they see the uber-wealthy getting huge rewards and even vast luxuries?"

"How do they feel? Frustrated. Mad as hell. The same way I felt when you stole my Halloween candy. Remember that? I got robbed. And you got candy you didn't deserve."

"Yeah, but I gave it back, didn't I? Most of it, anyway. . . . Now think again. We can see wealth—we can see the houses and the clothes and the yachts. But ask yourself this: do the rich see into Emily's home, into her kitchen, into her work? Do they *see* her every day on their screens? Do they see every day into the shelters where the homeless sleep?"

"Well, some of them might, sometimes. If they are seriously into doing charity. But not every day."

5. DOES IT MATTER?

"So now ask yourself: what's the effect of living in these different worlds?"

"It means people don't understand others and the way they live. And perhaps they don't give a damn."

"You've just defined the decline of community—the weakening of a shared sense of responsibility."

"Okay, smarty pants. Where are you going with this? Have a bunch of academics studied this decline of community?"

"*You* just came up with this idea in answering my questions, didn't you? And do you imagine that you're the only person to think of it?"

"All right then. So tell me. How do we know that this decline of community has occurred? And how do we know it's connected to inequality?"

"I want you to read a book called *The Spirit Level*. It's written by Richard Wilkinson and Kate Pickett. It was published in 2009, and there's been a huge amount of research since then that backs up their findings. The subject has a fancy name—social epidemiology—which means the distribution and causes of disease, and in this case, social disease or dysfunction. All this research shows that where inequality is greatest, social cohesion is weakest.

"In countries where inequality is greater, we see the following..."

- Fewer people agree that *most people can be trusted*.
- There is less spending on foreign aid.
- More people suffer from mental illnesses and more use illegal drugs.
- Infant mortality is higher and life expectancy is lower.
- There are more homicides and more suicides.
- Social mobility—moving into higher income levels—is lower.

"That's a horror show! And now I bet you're gonna tell me that inequality is to blame for climate change as well?"

"No, certainly not. But there are connections between inequality and climate change. Climate change likely makes inequality worse, at least in the short run, because the damage caused by climate change will affect low-income groups more and also racialized groups. And in

27

the other direction: which countries have gone the furthest in replacing carbon energy with clean energy? Usually the countries with low inequality. And we know something else—in the United States, there's an overlap between denial of climate change and rejection of redistribution policies. These two attitudes seem to be part of the same mindset."

"So what about politics? We live in a democracy. We have control over health and education and taxation, right? If we don't like what the politicians are doing, we can elect different ones."

"You're right. Canadians live in a democracy, and a relatively healthy one. But let me ask you a question. How much does our government lose in unpaid taxes because of tax evasion and tax avoidance?"

"I don't know. How could I know that?"

"But we had a federal election in 2019. Didn't you follow what was happening? What the politicians were saying?"

"Yes, I did! I wanted to hear what they had to say about the big issues. Like climate change and health care."

"So nobody told you that our government lost between $9.4 billion and $11.4 billion a year because of corporate tax evasion and tax avoidance. That's the estimate for 2014 by a reliable source—the Canada Revenue Agency."

"I don't remember hearing that."

"Do you know how much money that is? It's more than enough to end all homelessness and all poverty in Canada. So listen to this. In 1989, members of the House of Commons unanimously passed a motion to end all child poverty in Canada. And in 2009, the House of Commons passed a motion to 'develop an immediate plan to end poverty for all in Canada.' So how much progress has been made since then? During the election, what did the politicians tell you about their plan to end poverty?"

"I don't remember hearing anything. Was it an issue?"

"No it wasn't, except for anti-poverty groups. So tell me if you think there might be a relationship between democratic politics in Canada and economic inequality?"

"Sure. It's simple. Poor people don't vote, so they get ignored."

5. DOES IT MATTER?

"There's some truth to that, I think. But is it really so simple? It has always been true that poor people are less likely than others to vote. What I'm asking is this: is there some connection between *rising inequality* and *democratic politics*?"

"I don't know. There might be. If those who have an increasing share of income and wealth are also getting more influence or power."

"Now you're talking! But we have to remember not to be too swayed by what has happened in the United States. There it's much easier to show a relationship between politics and vast wealth at the top. It takes much more money today to succeed in politics than it did a few decades ago. The super-wealthy in the United States use their wealth to buy election results, to select politicians, to penetrate social media, to control messaging, and even—as many have argued—to undermine democracy itself.

"But we can't just look at the United States and heave a sigh of relief, can we? Who has the ear of the government in Canada? The anti-poverty groups? Or environmental organizations? Look at political lobbying. Between 2011 and 2018, the fossil fuel industry had 11,452 lobbying contacts with the federal government. An average of six contacts per working day. That's five times more than lobbying by environmental organizations. And that's just one example. Do you know that Facebook pays no corporate taxes in Canada? Netflix pays no sales tax (although there was a change in the 2021 federal budget)? Why not? Do you really think that this tax-free status is unrelated to the huge lobbying efforts of the technology sector?"

"Okay, that seems obvious. But why did you say that this story is not simple?"

"Because the connection between rising inequality and democratic politics is actually very complicated. For one thing, how can you separate the effect of inequality from other factors? Many people think there has been a terrible failure of political leadership in the United Kingdom, with Brexit and all that. If that is so, how is that connected to inequality, when so much else is going on?"

5. DOES IT MATTER?

"I don't know. You keep making things more complicated. Are you going to tell me to go and read articles from a bunch of political scientists?"

"That would be a very good idea! They aren't that hard to understand. And here, in a nutshell, is what they will tell you. Though there's no agreement on much of this, and the debate goes on."

- Higher levels of income inequality have the effect of depressing interest in politics and political participation.
- Voter turnout in Canada has been on a downward trend since the 1970s. There appears to be a relationship with rising inequality. In provinces where inequality is high, voter turnout is relatively low.
- Evidence from Europe suggests that the impact on voting varies among income groups, but in general, inequality tends to depress voter turnout and reduce participation.
- There is a mass of evidence showing that rising inequality is related to the rise of populism and *the revolt of the angry*. So traditional forms of political participation are at risk, and inequality gives rise to populist rage.

"Are we immune to such trends in Canada? I doubt it!"

"Okay, I'm convinced. It all sucks. You're not perfect, which I know better than anybody. But you're an encyclopedia with legs. So okay—inequality matters."

6. DESPAIR AND DEATH

"DO YOU REMEMBER our grandparents?" asked the professor.

"Sure I do. Especially Grandma, because she lived longer," replied the plumber.

"Sometimes I wonder what they'd think if they saw us now."

"They'd be okay. We've both done well, in our own ways. As for me, nothing wrong with being a plumber."

"I remember Grandma saying we could be whatever we wanted to be. And she said the same to us girls as she said to you guys. Live your dreams! Be what you want to be!"

"I bet they thought we'd be better off than they were."

"Yes, I guess so. And we are better off, in some ways, aren't we? I mean, we have a longer life expectancy than they did. And in other ways, we're better off. We probably have better diets. Some of us, anyway."

"Better off—but only in some ways?"

"Well, it's hard to compare. How many young plumbers in the 1960s would own a house? It wasn't easy then, and it's surely not easy now. I know you don't live in your house. You rent it out. But many of my colleagues working at the university don't own houses. Me? Well, I prefer not to. I'm on my own, and I don't have children—I don't need the space or the hassle."

"Are things getting worse?"

"Among young people today, I think expectations may be changing. The reality is that expectations will not be realized. Or not realized very easily. Even young professional people will struggle to own a house."

6. DESPAIR AND DEATH

"Why is this happening?"

"Our grandparents became adults just after the Second World War. It was a time of economic growth. The economy (real Gross Domestic Product) grew at a rate of around 5 percent a year in the 1950s and 1960s. We won't see growth like that again. And the important thing is the benefits were widespread. Of course some people benefitted more than others. But even if you were in the working class, you knew that you were better off than your parents and grandparents had been."

"An ordinary worker would not say that today."

"Probably not. Because growth has slowed down or stalled. And most experts predict that growth will remain stagnant for the foreseeable future."

"And there are times when economies actually shrink. Like during the pandemic in 2020. My plumbing business tanked—people were afraid to have us in their houses."

"Think about the effect of very slow growth. First, the fight over benefits—over who gets what—gets worse. People fight to protect their position. And they see the economy as a zero-sum game."

"The more somebody else gets, the less I get."

"Right. But it's more than that. Most people see their hopes crushed, their expectations dashed. Did your parents own a house? Do you think you can own a house? Not likely!"

"And there's all this talk about good jobs or bad jobs. But not everybody can get a good job."

"We see this effect among university students. They're under enormous pressure to succeed. Perfectionism has become an obsession. Going to college or university is not about learning any more—it's about getting a good job. And students know many will fail. These pressures have contributed to a serious mental health crisis among students. Stress, anxiety, depression, and suicide have increased rapidly in the last decade."

"Okay, so you're really good at telling stories about shit in everybody's fans. But what does that have to do with inequality?"

"Have you heard of *Deaths of Despair*?"

"No. What is it?"

"It's the title of a book by two American economists, Anne Case and Angus Deaton. Did you know that life expectancy in the United States *declined* for three years in a row after 2014?"

"Declined? I thought we were all living longer."

"Most of us are. But the trend was reversed in the United States. It was reversed in large part due to the rises in mortality among middle-aged white Americans. White, not Black Americans. Black Americans, of course, were already much worse off. But now there were many more alcohol-related deaths, deaths from drug overdoses, and suicides. Those are the so-called deaths of despair."

"What happened?"

"It's complicated. It has to do with the collapse of the white working class in the United States. With the loss of good jobs came crushed hopes and the loss of dignity and self-worth. Hence deaths of despair. And there's a connection to inequality—the upward redistribution of income and wealth toward the very rich."

"Has the same thing happened in Canada?"

"Not exactly. There are big differences in Canada, including our very different health care system. But you know about the opioid crisis, right? Nearly 14,000 deaths in Canada during the four years after 2015. Those were deaths of despair, of hopelessness."

"But those people were addicted to a killer drug. I don't get it. How can you connect such deaths to economic inequality?"

"Good question. You can't—at least not directly. You can't see a clear cause-and-effect relationship when you look at a single drug-related epidemic."

"You're getting amped, aren't you? I know you. When you sit up straight like that there's a mind bender on the way."

"Listen to me! Get your brain in gear and put your thinking cap on. Let's start with homeless people—those at the very bottom of the income scale. Their life expectancy is likely between 40 and 49 years. One study of Toronto in 2017 showed the median age of death being 48 years. So does that prove a connection between inequality and low life expectancy?"

"Yes of course. But umm . . . aren't there other reasons? Don't many homeless people suffer from poor health and addictions? Perhaps those things cause more deaths."

"That's exactly the problem! Death rates vary enormously in any population. And there are many causes, or conditions, of higher mortality. So how can one tell that inequality is a cause at all? And another thing: poor physical or mental health may come first—it may be a *cause* of having a poor job or low income—and so it may be a *cause* of inequality."

"Okay, let's give up. This is brain torture. I'm glad I'm a plumber."

"Don't give up yet! There are ways of sorting out these things to find out whether inequality is, or is not, one condition of both poor health and higher mortality—along with other conditions. We can ask: is inequality a *necessary* condition?"

"And what's the answer?"

"The answer isn't simple. Because inequality works through other conditions, like stress and depression. The answer is that there *is* a significant relationship between income inequality and mortality in Canada. Inequality is a condition of poor health, stress, depression, and psychological problems. These things contribute to alcohol and drug abuse and to tobacco smoking. And all of these things lead to premature deaths. But even in the absence of such negative behaviours, there is still a connection between inequality and mortality. Inequality kills. It kills large numbers of people. One study suggested it kills as many as 40,000 people a year in Canada."

"How come I never heard of this before? If you're right, where is this news buried?"

"It's not buried. But it's hard to see. Look at it this way. It's the same as if a Bombardier CS100 jet crashed every day, killing 110 people every time. If that happened, we would all notice. And that airplane would be banned for good. But deaths from inequality are slow. Harder to see. And even harder to see because there are many causes—not just one cause, like a plane crash.

"But inequality is still there. A silent but ruthless killer."

7. WE NEED INCENTIVES

"THAT SMELLS GOOD. What are you cooking?" asked the professor.

"Roast pumpkin and chorizo pizza," replied the plumber.

"Awesome!"

"Pour yourself some of that Burrowing Owl. It's quite good."

"For a beer drinker you seem to know a lot about wine. But how did you get interested in cooking? I even offered to teach you back in the day."

"Didn't I tell you? In my first year at BCIT. You were in Toronto doing your PhD. I lived in a house with a bunch of other students. One of them was in the cooking program. She showed me some of the ABCs."

"And you had an incentive, didn't you, you big lump. Your appetite!"

"That reminds me. You haven't said anything about incentives. Don't we all need incentives? And doesn't that mean there has to be inequality?"

"What *are* you talking about? And you complain that I get complicated! Tell me what you mean, in plain English."

"I mean a reward. An incentive is what gets people going—the best way is to offer a reward. And the best reward is money!"

"Okay. But we're talking about inequality, especially rising inequality. So this thing you call *incentive*—what's it got to do with inequality?"

"For a smart lady, you are sometimes a bit oblivious, aren't you? It's a no-brainer. People won't work unless they earn a reward—an incentive. So the rewards will differ. For more work, and especially more useful work, the rewards will be bigger. So there has to be inequality."

"So tell me what's wrong with what you just said. Go on! It's obvious what's wrong."

"There's nothing wrong! It makes perfect sense."

"It makes no sense at all. First, you've changed the subject. The subject here is *rising* inequality or *extreme* inequality. Why would we need bigger rewards for the top 1 percent and no increasing rewards for the rest of us? Are the top 1 percent so rarely motivated or so lazy that they need an extra million dollars a year to get them to continue working hard?"

"No, but don't we have to listen to them? They tell us that if their compensation is not competitive, or their taxes are too high, they will leave Canada and go to another country."

"And you believe them? Why? Compensation is much lower in Sweden and in Japan, and taxes on top incomes are a lot higher. Do you see the top 1 percent leaving those countries and moving to the United States? Headline: *Most Japanese CEOs move to the United States. Japanese economy crippled.*"

"Very funny."

"I don't think that threats of that kind are funny. Going on strike for better pay—when workers do it, it's really bad, but when capital threatens to strike, that's okay.

"Which leads me to the other reason why your *incentives* make no sense. You said that an incentive is a reward for work, especially for more useful work. But that's a fantasy. The *reward* that goes to the top 1 percent is not a reward for usefulness or productivity. Take large corporations, for instance. There's no relationship between levels of compensation and the productivity of companies or shareholder value or social utility (even if that could be measured). Economists have been telling us this for years. And what about the financial crisis in 2008 and 2009? It's well-known that CEOs who drove their companies into the ground still got big raises."

7. WE NEED INCENTIVES

"Okay, but this isn't a perfect world, is it? It's the principle that's important. We all respond to rewards or the promise of rewards for our actions."

"Too bad, my dear cousin! Your argument just collapsed. If, in the real world, reward is not related to effort or efficiency or productivity, then your incentive principle is a sham. So let's talk about where that *incentive* idea came from. It rests on a faulty assumption about human motivation. And that's what this is all about: motivation. What moves people to act.

"Motivation is complex and varied. 'Different strokes for different folks,' as the saying goes. This is true of the work world, just as it's true for the rest of life. So tell me. Why did you go to college?"

"I went because I needed to. To get the training I needed for my job."

"And you got the job you wanted?"

"Yeah. An apprenticeship first. Then I got my job three years ago."

"Why did you choose to train as a plumber?"

"Because I love doing it—always have. Remember? I was always good at fixing things."

"What do you find most satisfying about the work?"

"Doing a good job. Especially when I can solve a tough problem. Like last week when a lady had this busted pipe and ..."

"Okay, tell me some other time. Now answer me this: suppose I added 10 percent to your annual earnings. Would you work any harder?"

"No. I can't work any harder than I do now."

"Suppose I make another change to your work. Suppose I replace your current team with a few lazy guys who cut corners. And suppose I replace your boss with one who completely ignores you. Would you work as hard as you do now?"

"Probably not. I'd get out. I'd look for a job with another company. And I'd get one, too."

"So what moves you to work hard and effectively? The answer you just gave is this: seeing a job well done and working with a good team that shares the same goal. *These* are your rewards. And they are more important incentives than a pay increase. Though a pay increase would be good too."

"For sure. But I'm not sure that's true for others. Some people can get really greedy."

"Okay, what about very rich people? What about the CEOs who run big businesses? Do you think their motivation would be very different from yours?"

"How would I know? I don't know people like that."

"Their motivations have been studied. We know what they say because they have been surveyed and interviewed. Many of them, many times. They don't all say the same thing, but they do put income low on the list of motives. What do they really *want*? They want growth, they want sales, and they want really good contracts or projects that will make an impact. And they get satisfaction from building a good management team. From building strong loyalty among employees and from building a company that is firing on all cylinders. They value creativity—they are creators and builders. *These* are their rewards and their incentives.

"Many motivation experts argue that too much emphasis on money can be a *dis*incentive, not an incentive. Working for more money places the emphasis on external motivation—the pot of gold at the end of the rainbow. It distracts from intrinsic and social motivations and from satisfaction in doing the work itself."

"Why have I never heard this before?"

"Perhaps because in our culture we've been taught to sanctify money and wealth. But these ideas about motivation are well-known. And in some cultures, like Japan, they have learned these ideas very well. In Japan, CEO compensation is generally lower than for CEOs in the United States, and the ratio of CEO pay to employee pay is low. Why? There are several reasons. Japanese businesses give high priority to team work, company loyalty, and social motivation. The goal is *ikigai*—finding your purpose in life or reason for being. In business, that means uniting your passion and skill and satisfaction with the objectives of your company. And that's got little to do with personal financial rewards."

The rewards of social connection

"So let's just copy the Japanese!"

"No. We can't import Japanese culture. But we can learn the basic lesson. It's a serious mistake to equate incentives and motivation with financial rewards. The equation of huge financial rewards with incentives is a confidence trick. It's a scam that's easy to expose."

8. FIXING THE PROBLEM

"SO INEQUALITY IS a problem. Maybe it's a total shitstorm. But what can we do about it? If you're so smart, tell me how we solve the problem," said the plumber.

"That's a really big question. I think we need to begin by asking what a solution looks like. Don't you need to see your goal in order to figure out how to get there? So do we want perfect equality, where everybody gets the same income?" asked the professor.

"Well, I'd rather have the same income as you. What do you earn anyway?"

"A salary for an assistant professor can be between $79,000 and $98,000. I'm near the middle of that range. Salaries are lower in some academic fields than others, of course. And women professors like me still tend to get paid less than men, even when they have the same qualifications. But what do you think? Should everybody get the same income?"

"No, definitely not."

"Why not?"

"Well, first off, because that's not possible. There's no way that would work. Even in a communist society, some people earn more money than others. Also, we don't want perfect equality. That wouldn't be cool."

"So you've set up two criteria. First, a solution must have some chance of being realized. It must be feasible, even if it's hard to achieve. Second, it must meet some standard of desirability. Tell me, then, what might be desirable?"

"Okay, let's not get starry-eyed. Perhaps we want something that seems both possible and fair—like in those countries that have much lower inequality than Canada. Or why not what we used to have—like the lower inequality we had in Canada in the 1950s and 1960s?"

"Do you think such a change would require a revolution? Would it mean getting rid of capitalism, for instance?"

"Well, capitalism may be dying anyway. At least some people think so. But no, I don't think we would need to get rid of it completely."

"In other words, you could get your desirable equality through specific policies or reforms to the existing system. But tell me *who* would undertake these reforms? Should it be done by institutions: churches, professional associations, community groups? Or by businesses—those who employ people and pay wages and salaries? Or by governments?"

"I don't see how institutions could do it, though they might help. And employers? No, I don't see it. They would drop their gloves because higher wages would threaten their profits. It has to be done by governments. Perhaps both federal and provincial governments."

"Now you've arrived at the same starting point as the people who have thought about this a lot. There's a long list of such people. Here's a few of them: Thomas Piketty, a French economist; Anthony Atkinson, a British economist; and Joseph Stiglitz, an American economist. And in Canada there are several people, but I'll mention Lars Osberg, who has been studying inequality for decades. They all have suggestions about policies that could be enacted by governments."

"All right, but I'm not going to read their books. I do read books, you know, but some of their books are ginormous. Take that French guy, Piketty. First, he writes a book that's something like 500 or 600 pages long! And he was only getting started. But then he writes another book that's almost twice as long!"

"That's because he has a lot to say. Think of it this way. Piketty's books are like three books in one—you get three for the price of one! It might take longer to read one of his books, but that's because there's more in it!"

8. FIXING THE PROBLEM

"Are you his marketing manager? Don't give me the sales talk. And we don't need three books to tell us the solution. I bet I know what your solution is. Tax the rich!"

"Well, maybe. But think about it. Would that work?"

"It might. Government would have more coin. It could put more money into support for the unemployed and into housing for homeless people. And it could cut all taxes for low-income people. It could jack up the level below which people pay no income tax."

"Sorry, aren't you forgetting something? Put yourself in the position of a very rich person. If you found that your taxes were suddenly increased—and increased by a big amount—what would you do?"

"Umm . . . I suppose I would strong-arm the government to cut the tax increase."

"Yes, I think so. And even more than that. You would demand that your company give you a bigger pay increase to make up for the tax increase. And you'd tell your accountant to shift more of your money into tax-free accounts or even into a tax haven outside Canada. We know that this happens. When top marginal taxes increase, tax avoidance also increases. So you haven't done much to solve the problem of inequality, have you?"

"Maybe not. But no way it takes a shelf of books to tell us how to solve the problem. So don't give me another lecture. Tell me what we should do."

"Okay. Let me start with what Lars Osberg suggests. Then I'll add a few other ideas."

- **A guaranteed annual income.** Sometimes called UBI (universal basic income). Many Canadians already have this. Canadians over 65 get Old Age Security. There is also a Guaranteed Income Supplement for some people over 65. The UBI would expand the guarantee to cover all people of working age. The UBI is an old idea that gets a lot of support today because of the concern that robots and artificial intelligence may replace many traditional jobs. A UBI would also give financial security to people who do important service work—the jobs that robots cannot do so easily, such as caring for children or elderly people. This idea got a lot more attention because of the COVID pandemic in 2020.

- **A carbon price dividend.** Inequality is a serious obstacle to action on climate change. Carbon pricing or carbon fees (often, inaccurately, called a carbon tax) can help reduce greenhouse gas emissions. But to be really effective, the fees must be much higher than current levels. That would be a serious problem for people with lower incomes because they spend proportionately more on carbon-intensive goods. The idea is that all citizens would receive a carbon dividend to offset the fees, and the dividend would be much higher for those with low incomes.

- **Full employment.** Canadians have lived with high unemployment rates for so long that we have forgotten the benefits of full employment—benefits in terms of economic efficiency, productivity, and health. Full employment helps reduce poverty and sustain income levels. Bank of Canada interest rates should be designed to encourage low unemployment, not just low inflation. Government spending and taxation policies can give priority to job creation.

- **Fairer taxes.** A large number of informed observers, including many economists, are now urging that top marginal tax rates be much higher—closer to the levels of the 1960s and 1970s. There are many ways to do this, and it would require closing many tax loopholes and limiting tax avoidance, including transfers to tax havens abroad. Would the wealthy then leave Canada? The fear that they would do so has been seriously exaggerated. Cries of agony from the super-rich need to be put in context.

- **Higher wages.** We need higher minimum wages, and it has been shown that there is little connection between higher minimum wages and job losses (so we need to ignore the

8. FIXING THE PROBLEM

claims from right-wing think tanks whose "research" has been bought by business interests). Above all, the decline in trade union protection in Canada must stop. Inequality was lower in the third quarter of the 20th century, when the ratio of labour income to capital income was higher. That higher ratio must be restored, and this cannot be done unless a larger proportion of the work force is unionized.

"These are just a few of the main solutions that Lars Osberg proposes. But there are other proposals."

- **Inheritance taxes.** There are inheritance or bequest taxes in many countries but not in Canada. Why not? Of course people should be able to use accumulated savings or wealth to benefit their survivors or charities. But inheritance is a form of wealth transfer. Other transfers are taxed. Why not the inheritance transfer?

- **The wealth tax.** In his big book on *Capital and Ideology*, Thomas Piketty supports the idea of a graduated tax on property. Rich people can try to avoid taxes on income by shifting income around or using tax havens. It is harder to hide real property—and thus harder to escape taxation.

- **Financial transaction tax.** This is an old idea, and it already exists in many countries. The idea is to tax financial transfers, such as sales of stocks, shares, or currency. It is a kind of super sales tax, not on the sale of goods but on the sale of financial assets. An example is the Tobin tax—a tax on international financial transfers—named after James Tobin, an American economist who proposed such a tax in the 1970s.

"All right, you could go on forever! You're an encyclopedia. There's only one problem with your fancy solutions."
"What's that?"
"They're totally useless."

"What do you mean *useless?*"

"For a know-it-all university professor, you have no memory. We said that solutions should be desirable and feasible. Your proposals aren't feasible."

"Why not? If you adopted them, they would contribute to lower inequality. We know that! Because countries that do have such policies have lower inequality than we do."

"My dear cousin, you need to remember how things work in the real world. Here's an example. An example from my own business. If somebody has a leaky tap, what do they do? They hire me to fix it. They can't fix it themselves because they don't know how, and they don't have the tools. So I go, I remove the handle, I replace the cartridge, and I fix it. So you're saying that fixing inequality is the same. Get hold of the right tools, send in the right kind of economist with those tools, and fix it."

"I'm not sure that's a very good analogy."

"Why not? You're reducing a nightmare problem to a problem of missing tools. But missing gear is hardly the real problem! A few days ago, we agreed that rich people have huge political influence. They buy politicians, and they influence policies by lobbying. So there's no way they would let you introduce any of those fancy policies! Suppose I wanted to fix a leaky faucet, and I knew how. But I didn't have the tools. Why not? Because the people with the tools would not give them to me! Could I fix the leak?"

"Okay, I get your point. And so does Thomas Piketty, as a matter of fact. He says that policy is not enough. He says we all have to go back to democracy and rethink it. We even have to consider something he calls *participatory socialism.*"

"And that would mean grabbing control of the policy tools. But how? You're calling for some kind of revolution! Do you really think that low-income people are going to rise up and somehow grab hold of power? Most of them don't even vote!"

"Okay, so perhaps we have to start in some other way. And you're right—this is all about power. So ask yourself this: where does power come from? What power was it that brought down the Old Regime during the French Revolution? What brought down the monarchy and the entrenched power of aristocracy?"

"It was the guillotine."

"And what power is greater than that of the guillotine? Greater than the power of swords and cannons? What is mightier than the sword?"

"The pen is mightier than the sword."

"The pen. And what's that?"

"It means ideas, doesn't it?"

"Yes, ideas. Ideas put into words and actions. The pen is the enchanter's wand; the sorcery that can paralyze Caesar and strike the loud earth breathless. This is where power lives and where change begins."

9. WHAT WE VALUE

"YOU'RE MAKING MY head hurt with all your quotes from books I've never heard of. You're a bit of a show-off at times, aren't you?" remarked the plumber.

"Yeah, well, sorry. Maybe it goes with my territory. As for you, don't pretend to be a cynic," said the professor.

"Get off your high horse. What are you talking about anyway?"

"A cynic—a person who knows the price of everything and the value of nothing. That's according to a character in a play by Oscar Wilde, written in 1892."

"Are you telling me I'm a cynic?"

"No. I don't think you are, little buddy. Underneath all your bluster, I think you're an idealist."

"And you're a hilarious woman. You are! I'm not an idealist. I'm a realist."

"I think you're an idealist because I know you too well. I've been trying to get rid of you ever since we were children. And I'm betting I know your answers to the following questions. First, what is supply and demand?"

"That's economics. Supply is how much there is of something. And demand is how much something is wanted."

"Okay. More formally, the idea is that demand—how much people want—interacts with supply, which is a volume that is produced.

This interaction happens in a market. Eventually a balance, or equilibrium, is reached where supply equals demand. And in that equilibrium, the value of each unit of a good is determined."

"That's quite a story. But it's total crap."

"What do you mean it's crap? It's a basic principle of microeconomics."

"So you're telling me that we should listen to this story to decide what is valuable? And how *much* value there is in something? Heroin is worth a lot in the drug market. So it has a high value?"

"Ooops, sorry. I meant to say price, not value. The price of something is not the same as its value."

"Why did you say value, then? Was that a brain fart?"

"Yeah, I suppose it was. But it's easy to do, you see, because there's market price and there's something else—the value according to social or ethical standards. And it's easy to confuse these types of value. Anyway, isn't there a point here? If people, acting in a market, are prepared to pay a certain price for a good, and if manufacturers respond by producing that good, then don't we have one measure of the value of that good? A value given to it by consumers? You don't pay for something unless it has value to you—value in terms of usefulness or utility."

"You sneak. You're still trying to equate price with value. So let me ask you something else. Do people only buy things that will be of use to them?"

"Don't we have to assume that? We have to assume that people understand what they want. What's in their utility functions, as an economist might say."

"Okay, smart one, so tell me what the utility of heroin is?"

"Just because people want something doesn't mean it's really good or will benefit them."

"Okay, so people's preferences may be really wacko. They may prefer really bad things. So don't you have to ask where these preferences came from?"

"No, really! I don't need to ask that for the supply and demand model to work. The equilibrium price still comes from the interaction of supply and demand."

9. WHAT WE VALUE

"No, it doesn't. It comes from something before that. It comes from what consumers wanted and how they got their wants. What caused their wants in the first place? Their addictions, if they have any. Or advertising. The whole chain of events begins when suppliers amp up demand by telling consumers what they should really want. So your model is useless."

"No, it isn't! We have to go on to study those things that influence consumer tastes. That's another subject. But the model still works."

"Yeah? But only if you assume that consumers know what they want. . . . And I've had enough of all this! I came here for a nice dinner not for a lecture on economics."

"Hang in there, man! Ideas rule the world. Dead economists rule the world. Do you want them to get away with it? And you're no doofus. You can eat and think at the same time, right?"

"All right, lay off me. Tell me what other things I have to assume for your so-called economic model to be of any use."

"None, really. Well, of course we have to assume that there are markets in which competition occurs. Otherwise one producer could control the market and set a higher price than if there were competition."

"You've got to be kidding! So what about the market for gas for my car? It's obviously not competitive. The companies get together and fix prices. Is there a market at all?"

"Sure there is. The price is still related to supply and competition from producers around the world. And the price goes up in the summer when people drive more and are prepared to pay more."

"Right, but the companies still conspire to set the price in the winter and summer. Once the refined product gets to the gasoline pumps, there's no market. So go on—what are other assumptions?"

"We have to assume that the government does not intervene to influence prices."

"Oh, get away with you! That's Donald Trumpery. In the real world, the government imposes laws, standards, regulations—it's influencing prices all over the place."

"Okay, but here's the point. All economic models are tools of analysis. They tell us what might happen, when all other things are equal or held constant. It's called the *ceteris paribus* assumption—all other things being equal. Then you apply your model to data from the real world, and you can isolate those factors that don't fit. You can be much more precise about the factors that the model does *not* explain. Then you know what still needs to be explained."

"Umm . . . okay, that makes a little bit of sense. Despite your fancy words. But notice what you've said. All of the important and interesting questions appear when the model does *not* work. And the model tells you nothing about the value of anything. It doesn't even tell you how prices get determined. Which is what you said it did!"

"Yes, but you have to understand that the model is a starting point for understanding and analysis. It doesn't pretend to have a final answer on how prices get determined. The model remains a powerful tool for understanding."

"All right, I'll go along with that. I guess economists must learn all of this in first-year courses. But I'm still worried about something. You got muddled when you talked about price and value. And you tried to mix price with value. How often does that happen?"

"Not too often, we hope."

"Oh yeah? You told me you wanted to talk about supply and demand. And you gave me some damned homework, you tyrant! You told me to look at a textbook on economics. So I found it and read bits of it. It was really boring. Anyway, there was a chapter on *perfectly competitive markets*. And you know what? Nowhere did it say that perfectly competitive markets don't exist. Sure, it went on to talk about monopoly and oligopoly. But it started with perfect competition, and it said that perfect competition was a *benchmark* because only with perfect competition do you get full efficiency."

"So what's the problem with that?"

"It's obvious, book-brain! What does the word *benchmark* mean? It means a standard—something you use to judge quality or value. So you see? The textbook has snuck value back in without saying

9. WHAT WE VALUE

so. It sets up a *free* or *perfectly competitive market* as the standard for value. And any departure from that standard is of lesser value."

"Lesser value within a specific framework—the relations among production, distribution, and markets, and the standard of productive and allocative efficiency that is appropriate to that framework. There's a really important distinction here to be made between value, as determined by ethical standards and the desirable efficiencies in an economy. And economics deals with how things actually work. It studies the world as it is and not how it ought to be."

"Now I know for sure you're having me on! You're a really smart brainworker, so I know you don't really believe what you just said. First of all, you've been telling me that economics is *not* about how things actually work. It's about *models* of how things *might* work, so long as we accept certain assumptions. Second, you can't sneak away from the meaning of benchmark that way. That economics textbook was full of values and ethics. All over the place it was saying what *ought* to be."

"Okay! I plead guilty. I was setting you up, in a way. But I've just proven what I said at the beginning. You're an idealist. You may not know it, but you've been quoting an idealist—John Clark Murray—who was a Canadian philosopher."

"Never heard of him."

"He was writing in the 1880s. Like many philosophers at the time, he was writing about how economies work. He was also reacting to other writers—those who were seeking to find the *laws* that govern economies. The Law of Supply and Demand, for instance. Some people, Murray said, are claiming it's a law of nature and irresistible, like gravity. They are totally wrong, he said. And those people who say that any other law of economic motion is like a physical or natural law are also wrong. Such a *law* is nothing more than a tendency, and it may always be resisted or stopped by human will and agency. What's even worse, said Murray, is the way that economic writers tend to suggest that the natural law is also a moral law. That Supply and Demand tell people how they *ought* to behave. Every person who has any moral sensibility, said Murray, must condemn this thinking. The ideal—the moral law—comes first, and it cannot be deduced from an economic model."

"Okay, so I'm an idealist. So what? What does all of this have to do with inequality?"

"It has everything to do with inequality. First, because the kind of thinking that Murray exposed and condemned still survives. It allows many so-called neoliberals to argue that poverty and inequality cannot exist at all or that poverty and inequality are the inevitable outcomes of economic laws. And second, because we already decided that any solution to inequality requires more than just policies. It requires rethinking. We've got to go back to basic principles.

John Clark Murray: we are not the helpless spectators of a natural law

"So we begin by totally rejecting the idea that we have no choice—that markets always determine outcomes. John Clark Murray said it best . . .

"'We are not the helpless spectators of a natural law which dooms the mass of mankind to irremediable poverty, and accumulates the wealth of the world in the hands of the few.'"

10. EQUAL OPPORTUNITY

"SO, BOOK-BRAIN, YOU want me and everybody else to overhaul our mental gear and think differently?" questioned the plumber.

"I didn't quite say that, did I? What I mean is we need to think harder about what we already know," the professor replied.

"Okay, so I know that extreme inequality sucks. It has bad effects."

"And we can see that our economy tells us the price of everything and the value of nothing."

"So let me think while I finish this beer. . . . Here's what I think: if extreme inequality sucks, then we want to have a greater amount of equality. How's that?"

"Good start! But what do you mean by equality?"

"I mean that we all have more or less the same rights. And the same opportunities."

"The same rights? So that means we all have the same rights before the law, right? We all have the same right to justice, the right to act freely without disobeying laws, the right to equal treatment before the law, and so on. But how would that help reduce economic inequality?"

"Not much, I guess. The law can't forbid people from being rich or being poor."

"You're quoting the French writer Anatole France."

"Never heard of her."

"Him. He said, 'The majestic equality of the laws . . . forbids rich and poor alike to sleep under bridges, to beg in the streets, and to steal bread.' He was being sarcastic, of course."

"Okay, but what about opportunity? We want equality of opportunity. It's a really cool principle. There's always going to be some inequality, after all. The important thing is to make sure everybody has the opportunity to be successful. To succeed at whatever they want to do."

"Okay, that'll do for a start. But look at what you're doing with your *equality* principle. You're saying that equality is always *of* something: equality of rights or equality of opportunity. Equality never exists by itself. It always refers to some other good. That's why, many years ago, a really smart guy named Amartya Sen asked, 'Equality of *what*?'"

"Right on. Like equality of access to beer. Can I get another one?"

"Yeah, of course. So let's think about equality of opportunity. How would you ensure that such equality exists in society as a whole?"

"Education! Whatever you want to do, you need education. To be a plumber, I had to go to BCIT. So we need to make sure that everybody has an equal opportunity to get some education."

"That makes sense. Because we know that the education requirement for jobs has gone up in the last century. A long time ago, you could get a good job with just a high school education. Not today. And it's expensive! Today, if you go to university, it costs around $20,000 a year on average, if you live away from home. And the costs are going up."

"I'm never going to have children. Or if I do, only one. They cost too much."

"Yes, but education brings a big reward, in terms of lifetime earnings and jobs with lots of satisfaction. And another thing: do you know what the American Dream is?"

"Sort of. It's the idea Americans have that if you work hard, you can succeed. Anybody can get rich or even become the President."

"Right. And it's a dominant myth in the United States. It's an idea about their country that's widely shared. The only problem is that it's just a myth—it doesn't fit very well with reality. How do I know? Because social scientists have studied social mobility, including mobility across generations. They ask questions like: if a

10. EQUAL OPPORTUNITY

parent is in a specific income level, what are the chances that their children will be in the same income level—or a higher level or a lower one? The answer is that upward mobility is more likely in Canada than in the United States! Although there's a huge variation across Canada. Social mobility in Canada is closer to what you find in many European countries. And there's no doubt that our public education system and our high rates of post-secondary education help our social mobility. So if you want to live the American Dream, you are better off living in Canada!"

"Wait a minute! There's something wrong with what you just said. I'm trying to figure out what it is.... I mean, I'm not a professor like you, remember."

"Can you cut out the modesty please? You're a curious and astute thinker. So what's wrong with what I just said?"

"Shut up and let me rack my brain on this.... Okay, I think I've got it. You told me a while ago that inequality is lower in Canada than in the United States and lower still in certain European countries. But aren't those the same countries with lots of social mobility?"

"Yes, generally, but not always."

"Well, don't you see? The lower inequality could be what *caused* the greater social mobility!"

"And why might that be so?"

"You already know why, wonder-brain. Because more of them can afford to go to college or university. And their governments can put more money into schools. So if more equality *causes* more social mobility, then social mobility can't be the *cure*, can it?"

"Perhaps not! It might help, though. And a good education system is still important. But we were talking about equality of opportunity, weren't we? So what do you think about that principle?"

"Okay, let me think. Spreading out opportunity is a good thing because it allows people to move upward and maybe do better than their parents. But is there really such a thing as *equality of opportunity*, even in Canada?"

"Good. Keep going."

10. EQUAL OPPORTUNITY

"No, there isn't. Because if you're part of a low-income family, the *real* cost of going to college or university is going to be higher, even if there are student loans. The family would have to give up more. So the opportunity isn't really equal, compared to a richer family."

"Right! And another thing: the lower-income family is less likely to possess many books and other cultural things that work as motivators to get an education."

"Umm . . . so let me see if I've got this right. Equality of opportunity is a good thing. But by itself it's not enough. And everything depends on where you start from. Some people start from rotten places. It's like if I'm stuck with other people in a really bad place, and you tell us we all have an equal opportunity to escape, we should be happy. Because we all have an equal opportunity to escape! But we don't really—it's almost impossible to get out of the bad place with all its barriers."

"Now you're quoting T. Phillips Thompson."

"Never heard of him."

"A Canadian author and journalist. Grandfather of Pierre Berton, who wrote many popular books about Canadian history."

"So what did this Phillips guy say?"

"The prisoners in the Sing Sing prison have nothing to complain about. They all have an equal opportunity to climb the walls and run away. He was being sarcastic, of course. But he was saying that equality of opportunity is not a solution by itself. It can even be a deceit, a con job—a way of saying that inequality is not really a problem.

"Besides, suppose by some miracle you managed to create perfect equality of opportunity for everybody. There would still be inequality, wouldn't there? Those lucky enough to be born with exceptional talent would be super-rich. But is that just? As a very wise fellow named Michael J. Sandel asks, 'Do we deserve our talents?' And why should those who were born with lesser talents be condemned to poverty?"

"I dunno. But I have a huge talent for sleeping, and I'm totally whacked. I'm going home."

11. REWARDS AND MERIT

"I HEAR YOU won an award! Congratulations! So what was it for?" exclaimed the plumber.

"Just a book prize. For my book published last year," said the professor.

"That's great! So did you get some money?"

"No. Just a fancy piece of paper. Anyway, it's the recognition that counts. And I might get promoted this year. But I'm not sure I deserve the award."

"Well, why not? It's not like winning the lottery, which is just a matter of luck. You worked hard and did something awesome."

"Yeah, but so did others. It's really hard to decide why one book is more deserving than others."

"That's why there are judges. And I guess they're qualified to judge."

"So let me see if I understand what you're saying. There is such a thing as *desert*—which means being deserving of something. A person P is deserving of reward R because of accomplishment M. M has the quality of being worthy, and being worthy is the reason for the reward R. And in this case, it's the qualified judges who assess the amount of worth."

"Wow! Are you nit-picking or what? All I did was congratulate you for winning a prize!"

"I'm just trying to get things clear, little buddy. Because I want to understand what you're saying."

"Yeah, right, and I know what's coming next. You're going to connect this stuff about being deserving to inequality. Aren't you?"

11. REWARDS AND MERIT

"Yes. Well, maybe I won't connect it to inequality at first but instead to what is fair."

"And isn't it fair that people get what they deserve?"

"Yes, okay. So let's extend the principle from individuals to society. People P are deserving of rewards R because of worthy qualities or actions M. Let me ask you this: does this actually happen in society today?"

"Well, it might happen sometimes or even a lot. Some people get rewards that they've earned through effort or talent. That tennis player—Bianca Andreescu—was voted top Canadian athlete of the year in 2019. And she sure deserved that, didn't she! But look at how often some people get way more than they really deserve! And some people sure get what they *don't* deserve! Do homeless people deserve to be homeless?"

"Well then, perhaps the problem is that the principle of desert is not applied thoroughly enough. In a market-based system of rewards, too many people fall through the cracks, and reward is not always connected to being deserving. So you need to do some repairs to fix this disconnect. But what repairs?"

"I suppose you would have to figure out who is deserving and why. And then who is getting more than they deserve and who is getting less."

"Who? Or what? Is it the person who is deserving? Or is it the action or contribution that is deserving of either reward or punishment?"

"The person gets the reward or punishment because of what they did."

"And what if you cannot connect a contribution to a person? One of the greatest technological innovations in recent times is the internet. It was not created by a person. It was created by dozens of people and many institutions. And another thing: what do you mean by *what they did*? An action? Or contribution?"

"A social contribution deserves a reward—a contribution that is valued by society and benefits society."

"Measured how? Remember you can't use market values or prices when you're trying to repair the problems or gaps in the market system. So you need another scale of value. So tell me: how do you compare the social contribution of a newly composed opera with the contribution of a new clean-energy battery?"

"You can't. So perhaps being deserving means effort or energy and time. The more time and effort, the bigger the reward."

"Well, sorry, you think you can measure effort and energy? How would you do that? And even if you could record everybody's work hours accurately, would you then pay them in proportion to those hours?"

"Sounds like there's a problem with that. And I'm beginning to feel really sorry for your students."

"No need for that! And guess what? Your problems have only just begun. Think again about what you mean by *deserving*. You're saying: I deserve a reward because I was responsible for a good outcome or a valuable product."

"Well, of course! Why shouldn't I say that?"

"You can. But was the outcome a product from your mind alone? No. You produced a brilliant new clean-energy battery because you were born with a particular aptitude, and because you were born into a prosperous family that gave you science toys when you were a kid, and because your parents sent you to good schools and to university. So the valuable product was also a result of sheer brute luck. You aren't responsible for your luck, are you? So should you be rewarded for it?

"And another thing: no product of any value comes from a single mind or solitary effort. It's always the result of collective or co-operative actions. So can you say that you deserve the reward for a product created by many people?

"I could go on. But do you see my point? Your attempt to connect desert and reward fails. Your attempt may be based on a strong moral purpose, but it's just not feasible. There are too many problems."

"Too many problems again! You really are trying to mess with my mind, aren't you? Okay, so here's another idea. Let's get rid of these

11. REWARDS AND MERIT

problems about being deserving. Let's talk about qualifications. People should get rewards according to their qualifications. So to be a plumber, I had to get qualified—a few years studying at the trades college, a few years as an apprentice. That gave me knowledge and experience and skill. How long did it take you to become qualified for your job?"

"A four-year degree. One year in a Master's program. Five years in a PhD program. One year as a postdoctoral fellow. About eleven years."

"So there's a huge pile of knowledge and talent there. That's why you earn more than I do."

"Okay, great! You're looking for something that's clear and measurable. But I'm not sure you've succeeded. Here's one problem, for a start: can you really equate knowledge, experience, and talent to years of training? Why should years of training be the basis of just reward or just distribution? Or would you apply a rule of seniority—the more years in a job, the more you get paid?"

"No way! That would be seriously stupid."

"So let's broaden your idea. Let's talk about merit. It's an old idea. Aristotle said, 'Everyone agrees that justice . . . must be in accordance with some kind of merit.' But what exactly *is* merit? Merit has the advantage that it's not narrow, like years of training. It's broader. Merit means qualifications, experience, talent, proven ability—all those things that make a person worthy of the position they hold."

"Sounds good to me! So merit should be the basis of justice. And that means distributing rewards to people according to their merits."

"You know what? There are just as many problems with merit as there are with the idea of desert or being deserving. Qualifications, experience, talent, ability—how do you know what these *are* or where they exist? Many years ago, Amartya Sen pointed out, 'These things are means to ends.' You don't know what they are until you have a previous idea of their *ends*—the good results or products that you suppose they lead to. So you haven't gotten anywhere. All you've done is raise another huge question: what results or products have value, and what is your measure of value?"

"You sure make things complicated, don't you? As soon as I start to make sense, you tell me it's manure."

"Poor you! I'm sorry. But don't take it personally. Because it's not personal. And I'm not making things complicated. The complication lies in the idea or the concept."

"Okay, but I still think there's a point here. We all need to have some serious chinwag about what, or who, has merit and what is deserving. What about all those people who work in long-term care facilities for seniors? We started talking about them during the COVID crisis in 2020. We all saw the worth of what they were doing, and so we could see that they deserved more. They don't deserve to be overworked and underpaid."

"Okay, that's a good example. But why did you and so many others suddenly see more clearly that those workers are overworked and underpaid? And why did you think that something should be done about it? Because you saw two things more clearly. 1. The extremely challenging and sometimes risky nature of the work itself. 2. The health benefits of the work serving those in the care homes, and also serving those outside the care homes. You didn't need a concept of merit to see those things. Nor did you need any principle of desert or deservedness of people. What you needed was an understanding of the moral and social value of the work and its outcomes."

"I give up! This is a total mind warp. But I wonder why your book won that award. According to you, all the merit lies in the book. The book won the award! Nothing to do with you. You get no brownie points at all. I suppose if you had won some money, you wouldn't have accepted it. You would have given it to the book!"

"That's exactly what I would have done. Or rather I would have invested the money in my next project—not in some frivolous luxury. Because the prize is an investment in the work and its outcomes. And here I am pointing to another problem with the idea of merit. It confuses the person with the work. It equates human worth with specific deeds or accomplishments—education, qualifications, credentials. And the result is meritocracy."

"Meritocracy? What's that?"

11. REWARDS AND MERIT

"It means the rule of those who have merit. And it's a dream or rather, a nightmare. The term was given currency many years ago by an English scholar, Michael Young, who wrote a book called *The Rise of the Meritocracy*. Writing in 1958, he imagined a United Kingdom in 2034, where a new class of big shot meritocrats claimed the right to rule by virtue of their intelligence and declared the lower orders to be dunces and failures.

"'In the United States, meritocracy is the ideological excuse for wealth and inequality,' argues Daniel Markovits. Meritocracy creates a competition for credentials that only the rich can win. Then it adds insult to injury by telling the poor and less educated that they are failures. The exclusion of the uneducated and unmeritorious is really dangerous. It fuels anti-elitist resentment and populist rage.

"Of course we can all agree that people should be qualified for the jobs that they hold. But merit, when built into the self-justifying ideology of meritocracy, does nothing to solve the problem of inequality. It is a poisoned chalice."

12. PAY EQUITY

"STOP BANGING ON the door! Come on in!" shouted the professor.

"I need to pee," yelped the plumber.

"You have such a weak bladder."

"Yeah, yeah … get me a beer."

"No way, little buddy. I'll pour you a coffee."

"I just had a few after work."

"Sit down, drink your coffee, and I'll get you something to eat."

"Okay, smart lady. You're so pure—you never drink too much."

"I'm too busy."

"And you don't seem to have any social life, if you know what I mean. For a wizard-brain desk jockey, you've got a lot going for you."

"I don't need an earbashing from you, you big lout."

"So what's your type anyway? Some ancient, dead Greek?"

"Definitely not a puffed up Alcibiades like you. Anyway, park your assumptions at the door. As you know, I don't walk the same paths as you."

"Nah! You love those dead Greek guys."

"Alcibiades lived in Athens. In one of Plato's dialogues, he showed up drunk at a dinner party and insulted Socrates."

"Yeah, well, Jeannie's wondering why I spend so much time over here. She's getting jealous."

"Well, bring her with you next time! We get along fine. I like her. You know that."

"All right, all right. But she's not into philosophy. And what's my love life got to do with inequality anyway?"

"I wasn't talking about your love life."

"Equality of the sexes! Pay equity! I've heard of that!"

"Don't shout! But maybe you have a point, no credit to you. There's a gender wage gap. For every dollar made by men in Canada, women earn about 87 cents. And look at the glass ceilings! In 2019, one study showed that among 1,200 company executives in Canada, women earned 68 cents for every dollar earned by men. And of course there are way fewer women in top positions than men. And look at politics! In the 2019 election, only 29 percent of those elected to the House of Commons were women. The participation of women in the labour force has been rising, but it is still below the level of men—and women are twice as likely to be working part-time."

"There you go! Problem solved! Get more women into jobs and pay them the same! Inequality all gone."

"Yes, employment and pay equity are absolutely necessary, and it's fair. But there's a problem."

"Yeah, yeah. Always a problem. Always a problem."

"Listen to me, you slug. This is a huge issue, and we'll need to come back to it later. But for a start, let's suppose we decided to get rid of gender inequalities in the economy by ensuring equal numbers of women and men in most jobs. Suppose, by some miracle, there were equal numbers of women and men among the top 1 percent—among executives and bankers and derivative traders and university presidents and so on. But then there was no other change. And suppose that in low-paid jobs—like service sector jobs—there were just as many men as women. Impose gender equality in this way, and pay equity would follow, right? But the gap between the top 1 percent and the rest wouldn't have changed. The gap between CEO pay and the pay of employees would be the same, wouldn't it? There would just be a different mix of people in the same income levels! So there would be no change to the structure of inequality, would there? Would there? . . . Hey, you! You can sleep on my couch if you want, but don't snore!"

13. WHAT IS PROPERTY?

"HOW IS MOTHER?" asked the professor.

"She's okay. You realize it's a year now since Dad passed away," said the plumber.

"Is she going to stay at the house?"

"She hasn't decided. And I fixed a few things for her a while ago—a few leaks, rotten fence posts. But I really think she should move into a smaller place."

"Yeah, I agree. And I've talked to her about it too. Though I would sure miss the old place. It means so much. The place that became my real home.... But she'd do all right if she sold, wouldn't she? It must be worth a lot."

"Yeah. Even though property values have dropped, it's worth a lot."

"So what do you mean by property values?"

"Oh my God, here we go again. I mean the amount of money she'd get! The selling price!"

"Okay, so the selling price is one measure of the value of the property. But what do you mean by property?"

"The house! And the land!"

"Okay, that's true in this case. But does property always mean house and land? Surely houses and land are just specific examples of property. So property is something bigger."

"Yeah, okay. My guitar is my property. So are my tools. So property is anything I own."

"So property means ownership or the right of ownership over some good or asset. And the asset doesn't have to be a physical object, does it? It could be a cultural thing, like a song. You could have the right of ownership over a song that you've composed."

"That makes sense. But wait a minute. Are you saying that property is the same as ownership?"

"Not quite. Ask yourself: is it possible for there to be property where the good or asset is *not* owned by anybody?'

"Umm... not sure. Maybe. What about land that is owned by the public or by the government? Like parks?"

"Yes. Parks have an owner—a collective owner. The right of property is not individual and not private. The right is one of access. And is there any land, anywhere, that has no owner?"

"I doubt it!"

"Never heard of Bir Tawil? It's unclaimed territory between Egypt and Sudan. Yes, there are small bits of land on earth that are unclaimed—they don't have an owner. There used to be many more of such lands. They are often referred to as common property—property where the right is common to all. The key here is the right of access. All members of the community have the right to access and use the property."

"Okay. So what? What does all this have to do with inequality?"

"It's got everything to do with inequality. Because this is where serious thinking must begin. We live in a time when private property has become sacred. Some people—like Thomas Piketty—call it proprietarianism. That means treating private property as sacred, as a supreme value, as the source of individual freedom. So we must rethink property and its meaning. And more than that: it may well be that *inequality* begins when property becomes a *private* possession—when property becomes a bundle of rights securing personal and exclusive control of assets (mainly land)."

"When did that happen?"

"Have you ever heard of the English philosopher John Locke? He wrote about property in the late 1600s. He thought that God gave the earth to all human beings in common. But eventually there came private rights or entitlement to bits of the earth, and with that

13. WHAT IS PROPERTY?

move from common to private came, 'disproportionate and unequal possession of the earth.'"

"Okay, but wasn't that inevitable?"

"Locke thought so, or at least he thought that private possession was a good thing. But nothing is inevitable. All outcomes are avoidable because they are the result of choices between alternatives. And for thousands of years throughout human history, most property was common property."

"Well, I like private ownership. I own a piece of land and a house, and that gives me freedom. I can use it to charge rent and get money. Or I can sell it if I want."

"How much freedom? Can you do anything you want with your land? Can you demolish the house and build a skyscraper?"

"Of course not! There are zoning bylaws."

"So your right of ownership is a limited right. And why is it limited?"

"I guess because there are some things you just shouldn't do. Like, in my neighbourhood, nobody wants skyscrapers. It's a residential neighbourhood. And nobody wants a bar or a night club next to their house."

"So your right is limited by community values. Values that are upheld by governments through laws and regulations. So is private property always a limited right?"

"Perhaps. But with some private properties, maybe there aren't many limits. Isn't this why things sometimes get nasty? Some big oil companies have a right to pump crude oil out of the ground. They own the stuff—or they have rights to it—and they sell it. But their right is limited by environmental regulations, and then many people argue that their right should be even more limited. Then things get nasty."

"Set aside the controversy for a moment. Beneath it there's a principle. So what principle are you and others applying? Not just to oil but to private property in the economy?"

"A principle? How about this—private property, including capitalist property, is okay, so long as it serves human beings and so long as the purpose is to benefit humanity as a whole."

"Now you're quoting Thomas Aquinas."

"Yeah, sure! And I read him every night before saying my prayers. In Latin."

"I'm serious. He's the Catholic theologian who lived eight centuries ago. Everybody who thinks seriously about property refers to Aquinas. He said, 'God has dominion over all things.' If you like, the original owner is God. Nature, or natural law, tells human beings to make use of material things in order to survive. Next, human reason tells human beings to make rules for private property because reason tells us that there are benefits when we allow private rights over material things. Therefore, private property is a tool. It is a means to an end—and the end is human flourishing or enhancing human life. Thus the right of ownership is never absolute. And if the end of ownership is not human benefit or human flourishing, then the right does not exist.

"So what follows from that? One thing is this: private ownership is no longer lawful when it causes others who are in need to be denied access to necessary resources. Also, if a person is in desperate need and has no other alternative, then taking a material means of survival from another person is not a sin. It's not theft!

"You get the point. Private ownership is not natural—it's *reasonable*, and the right is very limited. It's always subject to a higher justice—a universal purpose, or what we might call the common good."

"Okay, I can agree to that. But what does that all mean for us, down here in the real world?"

"Obviously Aquinas lived in a world that was very different from ours. And some of the things he said sound weird to us. But if we think about what he said, it helps make our own ideas sound weird. It gives us a new way of seeing our world with fresh eyes. It helps us think outside the box. So take those principles that I quoted from Aquinas. What do you see? Apply the principles to oil companies, for a start."

"Umm . . . that's not easy. And we might disagree, but the oil companies would have to make pretty good arguments about general human benefit and flourishing. The short-run gains in terms of jobs and profits would not be enough."

"That's a start. And yes, we might disagree, but there's no conversation more boring than the one where everybody agrees. That's another quote, by the way. Sorry, go on. Apply your principles to inequality."

"Okay. Where there's a surplus—that is, a big pile of property or wealth that clearly does not serve the common good—there is no right of possession. The possession of a surplus is not lawful or just."

"Can you think of an example?"

"I heard about this company that wants to sell return flights to the moon for 100 million dollars. And what about those people who buy up houses—big mansions—that nobody lives in? I don't see any common good there."

"So the principle helps you *see*. Does it help you see what actions we might take?"

"Well, we can't deny people the right to fly to the moon if they want to. But they can do that only after they've done what's right. They can't claim to be serving any common good. Tourism of that kind is just for personal pleasure. So the money they spend is not theirs to spend—not all of it, anyway. It's community property."

"Yes. Because all property is created by society, as a social right that's serving the social good. Everything else follows from there."

14. WHAT IS FAIR?

"I BET YOUR students are smarter than I am," said the plumber.

"No way! You're different, though. You're always ready for a disquisition," said the professor.

"A what?"

"A disquisition. A discussion or discourse."

"Yeah, I can be ornery. But I didn't do that well in school."

"You do okay for a stunned bunny who drinks too much beer."

"Speaking of which . . ."

"Okay, I'll get you one. By the way, I hope you're remembering to take your vitamin D."

"I don't need to remember. I've got you to remind me."

"So what do you want to talk about?"

"Well, we've figured out a few things. And I can see why we should think about property. But there's something missing. Inequality—it's always been there. It may have gotten worse recently, but it will always be there, won't it? So there's no such thing as a perfect solution, is there?"

"I think you're right. There's never a perfect solution in life. Did you and I ever find one? . . . Anyway, where does that lead us?"

"Well, maybe what we want to agree on is some principle about when inequality is okay and when it's acceptable."

"As I was saying, for a stunned bunny who drinks too much beer, you're doing okay. So what you're saying is you want to talk about John Rawls."

"Who?"

"John Rawls. American philosopher. Very famous. He wrote a book called *A Theory of Justice*. Published in 1971. Philosophers have been talking about his ideas ever since."

"So did he figure out when—or what—inequality was acceptable?"

"He sure tried. And maybe he did better than anybody else in the 20th century."

"Okay, so he's a hard-thinking dude. Tell me what he said."

"Not sure I can. You should read his book. It's impossible to summarize."

"Hey, cousin! You can do the impossible! I know you can. You've got more brain cells in your little finger than I've got in my head."

"Stunned, and self-deprecating too. Maybe I beat up on you too much when we were kids. Girls rule!"

"Well, at least this guy Rawls was around not so long ago. Like, not 800 years ago."

"True. So let me ask you: if you had to design society all over again, from scratch, how would you do it? You want a society in which fairness and justice prevail."

"Wow! I don't know. I guess I'd have to figure out how to define equality. And how to distribute income and wealth in some way that is fair—or as fair as possible—to all."

"But right away you've got a problem. Your idea of what is fair is going to be very different from other peoples' ideas. You're a whitish, Anglo-Saxon, Canadian guy, who's 28 years old, with a job and a thirst for beer. Won't all that affect your judgement? Your design for society will reflect your bias!"

"I suppose. So what can I do about that?"

"You can accept that when you design society from scratch, you're in a thought experiment. And in that experiment, you don't know who you are! You must stand behind what is called a veil of ignorance. You don't know your sex, race, nationality, your conception of good, or your personal tastes."

"Wow! Not sure I can do that."

14. WHAT IS FAIR?

"Well, try! Now, says John Rawls, we can start thinking about fairness in society. And do you think that *justice as fairness* can be stated in one simple principle?"

"I doubt it. So what does he give us? A bunch of principles?"

"Sort of. Think of it as a matrix or a combination. Each principle in the combination is necessary, and some take precedence over others."

"My head hurts already. Go slow!"

"First principle: each person has the same claim to basic rights and liberties. The scheme of rights and liberties applies to all. Freedom of conscience, freedom of speech, the right to vote, the right to hold political office, the rule of law—the rights and liberties of a democratic society."

"That's a good start."

"Second principle: social and economic inequality may exist. But—and this is a big *But*—inequality may exist only if two conditions are met. The first condition is: all offices, all positions, all jobs are open to all under the condition of fair equality of opportunity."

"Hang on! There's a problem there! We talked about equality of opportunity already. How do we know it's not a con job?"

"That's why you need to read his book. The devil is in the details, as they say. Rawls says that the class or income level into which you were born should not affect your chances of ending up in any particular occupation or income level. It's your talents—and your willingness to use your talents—that determine your outcome."

"Okay, let's go with that for a moment. But I'm still wondering how that kind of equality of opportunity can be a prior thing, when it can only be the *result* of a really high level of equality that's already in place."

"Second condition to the second principle: and this is where we get to what you wanted—a principle that tells us when inequality is acceptable. *Inequalities are to be of greatest benefit to the least advantaged members of society.*"

"Wow! Let me see if I get this straight. Inequalities are okay—they are fair—so long as they are of most benefit not to the rich, but to the poorest. The least advantaged. Inequalities are okay if they make the worst off better off."

"Right. So what do you think?"

"I don't know. Sounds okay. But how can an *increase* in inequality work to the *greater* benefit of the poorest? That sure doesn't seem likely."

"Maybe it's not likely. If so, that means it's a really tough test. The rich can get more, but only if those with low incomes get even more. The rich might double their incomes, but only if the poor can more than double their incomes."

"You know what? You have to assume something, don't you? You have to assume that when there's economic growth—a bigger pie—then a larger share of that pie will end up in the hands of poorer people."

"Yes, but doesn't that just confirm that it's a tough test? And if society or the government were to apply such a test, it would mean that the rich would have to pay a lot more attention to how the poor were benefitting from this growth. They would have to share the benefits of growth!"

"Okay, I see that. But there's something I don't like about this. Something's bugging me. I mean, you start by saying that inequality is okay, so long as the poor benefit, even if they benefit only a little bit. So a rich person could get an extra million dollars, as long as a poor person gets an extra $10,000. The gap may have narrowed a bit, but it's still huge."

"You're on the right track, or at least one track. Because Rawls set off an explosion—an intellectual explosion—his principles, and the rest of what he wrote, were brain candy for the hard-thinking types. And people are still arguing about what he said.

"Anyway, here's why you may have a point. I know you hate numbers but look at this table."

"What am I looking at?"

Brain candy for the hard-thinking types

Economy	Lowest third	Middle third	Top third
A	$20,000	$50,000	$100,000
B	$30,000	$100,000	$500,000
C	$25,000	$80,000	$300,000

"Average incomes of people in three different income levels. In three different economies. You can see right away that economy B is the richest—it has the highest income levels overall. But apply Rawls's principles. Which economy is to be preferred?"

"I suppose it's B because the poorest are the best off. If that's what you mean by the greatest benefit to the least advantaged."

"Okay, *if* that is what I mean. But look again. Which economy is most unequal?"

"Uh . . . let me figure this out. Well, it has to be B! Uh oh! The best economy might be the one with the greatest inequality! So principles that were supposed to impose fairness ended up permitting inequality!"

"Perhaps. I say *perhaps* because the debates only begin here. And there are some who insist that Rawls was far more in favour of equality—far more egalitarian—than your doubts might suggest."

"I know I started all this when I said we should think about when inequality is okay or fair. Now I'm not so sure. Inequality is okay *if*. . . that sounds like a negative approach. Perhaps we should say: more equality is better *because* . . ."

"Let's have another bowl of ice cream. We've earned it."

15. WELFARE STATES

"COME ON IN. Take a load off. You look tired," said the professor.

"Bad week. Gimme a beer," replied the plumber.

"What happened?"

"Lots . . . a horror show today. Chronic backups and a stink of sewage in an old house. Turns out that the vitrified clay pipes were totally bunged up with tree roots. We looked at it with a Gen-Eye camera. Never seen such root balls. And the owner had built a huge deck over the pipes! So we couldn't just dig it up. We've got to use a heavy-duty pipe bursting system. And the owner was really pissed off—he thought we could fix it in an hour or two!"

"A plumber fixes the problem of leaky pipes. A philosopher tries to fix leaky arguments. I bet the plumber is more successful."

"And then one of our best guys quit. Moving to Kelowna. Then we lost two contracts we were sure we would get. A shitty week."

"I'd say you're in a condition of illfare."

"What?"

"Illfare. Faring badly. Not being well off."

Ill fares the land, to hastening ills a prey
Where wealth accumulates and men decay.

"There's a bit of poetry for you. So illfare is the opposite of welfare or well-being."

"As I said, shit happens."

"Okay, so what is *welfare* then? And how do you know it when you see it?"

"Might be very different in your ivory tower than in my trade."

"Yes, because the things we do are so different. But still, can't we say that there is such a thing as satisfaction, even if the sources of satisfaction differ? And some things give lots of satisfaction, while other things give little or none?"

"Yeah, I guess so."

"So each of us can see an amount of usefulness or satisfaction that we might get from any good or service."

"And right now I could get a lot of satisfaction from another beer."

"Okay, mopey man, I'll get you one. If you promise to listen to what I'm saying! So you get satisfaction from a beer, but can you really say how much? Can you measure the amount of satisfaction so that you can compare the satisfaction from a beer with that of listening to a Bruce Springsteen song?"

"No way. That would be seriously dumb."

"This problem—can we add up satisfactions?—became really urgent more than a century ago. It became a real brainteaser when economists started to think about how to solve problems of poverty and inequality. In those days, many economists were trying to figure out how to maximize satisfactions in society as a whole. And they were mixing up satisfaction and utility. Utility meant usefulness in terms of pleasure or satisfaction. So the new challenge was this: how does a society maximize utilities? It was sort of like saying that we want the greatest happiness of the greatest number of people. That's the old utility formula."

"That sounds really neat. But I can see the problem. Can you measure these things? And what *is* happiness anyway?"

"I don't know. These are huge problems! And these doozies gave birth to a whole field of economics—welfare economics. It began more than a century ago. The big issue was: is it possible to maximize welfare, either for individuals or for society as a whole? And what would that mean?"

"Welfare—doesn't that mean well-being? Being in a good state of health and having decent living conditions?"

"Yes, but you still have to define those terms. And don't you need measures of such things? Eventually economists developed

15. WELFARE STATES

a new understanding of utility. It wasn't just satisfaction. Utility refers to the preferences that an individual has when faced with a set of choices. Economists started to talk about utility functions. A utility function shows the preferences, or order of preferences, that maximize preferred outcomes. The utility function is a way of demonstrating the combination of preference choices most likely to maximize satisfaction."

"Do all people try to maximize their utilities?"

"Yes, one would assume. And this relates to welfare. Society's maximum welfare is achieved when you maximize the sum of individual utilities."

"Wow! That would still be hard to do, wouldn't it? A plumbing business can add up the money received from clients over a month and get total revenues before costs. But that's because the thing we're adding is clear and simple: dollars. Can you add up utilities?"

"Not easily. But at least you can start thinking creatively about such things as overall welfare and social efficiency, rather than just growth. And you can start thinking about rules and tests. So here's one example: it's an efficiency test. A situation is best only when no individuals can be made better off without making somebody else worse off. This rule has a name—it's called Pareto efficiency."

"Umm ... I'll have to think about that one. It sounds simple, but maybe it isn't. Anyway, does all this have something to do with the welfare state?"

"For sure. The welfare state appeared gradually, as one solution, in the world of politics and government. But what do you mean by welfare state?"

"Things like medicare, maybe. Employment insurance, for sure. Income for seniors."

"And social assistance for very poor people. And income support for people with disabilities. And some kind of benefits for families with children. In Canada, we used to have a universal family allowance—an amount of money paid every month to mothers for each child in her family. In 1992, this was replaced by a Child Tax Credit."

"Okay, so we have a welfare state. So what?"

"So what do you think this welfare state is *for*? What's the purpose?"

"I guess the idea is to support people who need help. If they get sick, or if they are unemployed, or they can't work, they get assistance."

"Is that all? Does the system work so as to maximize the sum of welfare or well-being? Is it supposed to maximize the sum of utilities?"

"Well, I suppose it might. But you could never know for sure, could you? And isn't there a problem when you add up things? You might have a lot of people who are really well off, with their utilities over-maximized, and a small number who are starving. The sum could still be big. But that would be a crappy system."

"So might there be another goal?"

"Yes. To minimize poverty. To keep people from falling into real hardship."

"You might call it maintenance—to maintain a certain minimum standard of living. And to use economic growth to make sure people don't fall through the cracks. A rising tide will lift all boats, as the saying goes."

"And why the hell doesn't it work? Why are there so many homeless people in Canada? Why do so many people go to food banks?"

"Good question! The simple answer is that the welfare state, or social security system, is really quite small in Canada. And it has declined in recent decades. And it has many loopholes."

"Okay, but what does all this have to do with inequality?"

"You tell me. I bet you know the answer from what you've said already."

"Well, if welfare does provide some basic support for those who need it—and if it doesn't always work—then it can't do much to reduce inequality, can it?"

"And from what you know, does the system say anything about *how much* should be transferred from the rich to support those in need?"

"Don't think so. But the rich support the system by paying taxes, don't they?"

A rising tide lifts all boats?

"Of course. But everybody pays taxes. Even if you don't pay income taxes, because your income is so low, you still pay taxes, such as sales taxes. So everybody supports the welfare system. We all pay for it, including those who benefit from it."

"But wouldn't inequality be worse if there were no welfare state?"

"The *effects* would certainly be worse for those who need support the most. But the ups and downs in inequality are largely separate from the welfare state in Canada, if not elsewhere. When did inequality decline most rapidly? Before the welfare state was created. When did inequality start increasing? In the 1980s when the welfare system was still relatively strong."

"Are you saying that the welfare state is useless?"

"Not at all. It's essential. Not just for those who benefit directly, but for all of us. But there's a big downside, I think."

"What's that?"

"Most Canadians don't benefit directly from most of the programs (except medicare of course, if you include that in the welfare state). So they don't see the importance of it all. And they often think the programs are bigger than they really are. The result is complacency. And even an excuse. I would call it The Big Excuse: inequality does not matter much because in Canada we look after the poor."

"And I came here so you could cheer me up! You haven't done anything for my illfare. You buzzkiller."

16. EQUALITY AND CAPABILITY

"I GOT THE pizza. The kind you like. As well as my pepperoni," said the plumber.

"Great! What do I owe you?" asked the professor.

"Nothing. I've been drinking your beer. Like, regularly. We've been getting together for dinner a lot—like every two weeks or so."

"True! So pizza for beer. Fair enough, I suppose."

"Fair . . . fairness . . . weren't we talking about that a while ago?"

"Yes, we were. Hard to know what it means, though. You know it when you see it, and you can sense when it's not there. But what is it exactly?"

"Hmmm . . . I suppose it means we've spent roughly the same amount on these things. So there's a rough equality in what was spent."

"Sure. In this case, we both give, and we both receive. There's an exchange. The things we exchange are different—there's respect for our different preferences—but in the exchange, there's an equality that we both accept. But is that the whole story? The whole story of the equality that may or may not exist between you and me?"

"No, of course not. I treat you as an equal, don't I?"

"We're cousins—and we've been like siblings since we were children. So the question of equality does not arise. As Aristotle said about friendship, 'Between friends there is no need for justice . . . indeed friendship is considered to be justice in the fullest sense.' We could replace justice with equality, couldn't we?"

"Yeah, I suppose."

"So there's much more to it than making things fair, or squaring things, when I provide beer and you provide pizza. So what else is there?"

"Well, you didn't rat on me when that log caught fire."

"You mean when you stole those cigarettes? And you and your friends smoked them inside that big hollow log, and it caught fire?"

"Yeah. And you got us out of there just before the fire department arrived. That was a real bailout."

"Even though you boys were total idiots. But that kind of thing is easier, perhaps, because we're close. We were thrown together. Our parents left us on our own a lot, for obvious reasons. We know what alcohol did to them."

"Yeah. And I always thought you were one of the boys."

"That's because if we got into a fight, I always won. But that raises a question. Does having respect for a person—or for people—of the opposite gender mean treating them the same? As though they were the same as you?"

"Not really. I guess I knew you and I were different in lots of ways."

"Here's an example. What does equality mean in sports? Does it mean that girls and women get to play on the same team as boys and men? Or does it mean that they play on their own teams, but get equal support? And if it's a professional sport, they get equal pay."

"Sometimes girls get to play on boys' teams. There was a girl on my high school basketball team. She was amazing. But most often, girls play on girls' teams and boys play on boys' teams."

"So what's the principle behind this?"

"A principle? Maybe equality means giving people the same rights, the same access, and comparable rewards, while respecting their differences."

"Okay! But notice how complicated this is getting. Notice how much you've put into the equality box. Rights, access, rewards, along with respect for differences as an essential condition."

"Yeah, I guess it's kind of complicated. Too much for my small brain."

"No! Not at all. Think of it this way. Equality is a big box, a very deep one, with lots of possible values and meanings in it. So it's very hard to know what we're referring to when we use the word. Even

16. EQUALITY AND CAPABILITY

when we think it's a fundamental value. But we have to know what we mean. Especially if we want to say that the solution to inequality requires a greater degree of equality."

"Makes sense. And a greater level of equality can't be measured simply in terms of dollars or income."

"So what are we trying to equalize, to a greater degree?"

"Well, I said rights, access, and rewards. You said we were getting complicated. Perhaps we just can't simplify."

"Well, is it possible to come up with a general principle? That might be simplifying, in a way, but it would be energizing. It could inspire agreement and action. So here's an example: in Canada, we talk about equality among provinces. And we have something called equalization payments. Does that mean we are trying to treat provinces as though they are all the same?"

"Sure! We want to have the same level of education and health care in all provinces."

"Not quite. First, the goal is a comparable or similar level of public goods in terms of quality. Second, the principle about respect for differences applies. To say that provinces are equal in specific ways doesn't mean they are the same. It means respecting differences. Thus in Quebec, there's a different kind of law, there's a different pension plan, and there are different language rights. That's just one example. So equality means accepting and even supporting differences."

"Makes sense. So provincial governments can operate in different ways. They can try to support different values and even distinctness—is that a word? But they all have a more or less equal ability to do that."

"Yes! So let me be a professor and suggest a principle. A principle of general application. Equality of what? Equality of functionings or capabilities. Functionings are things we can do or be, like being well-fed or being literate or voting in elections. Some functionings can be more complex, like taking part in a community. Capabilities are the actual or real opportunities one has to achieve functionings. Think of capabilities as effective or real capacities to do or to be, and hence to realize a valued state of being. The measure of justice is the presence of such capabilities. Justice exists where capacities exist for all, or for as many as possible."

"Well, bash my brain, will you! That all sounds kind of rarefied to me. So why should we try to think this way?"

"It's called the *capabilities approach*. It was developed by the economist Amartya Sen and philosopher Martha Nussbaum. The advantage is that it gets you out of the problems connected with the welfare approach or with the idea of utility. It allows for a wide variety of actions and goals, without having to say we do things just because they yield utility. It de-emphasizes resources; it says we should not focus on trying to equalize income or specific material things. And it allows huge respect for diversity, including diversities of race, location, and context. And here's another advantage: it leaves lots of room for responsibility and choice. That is, if you focus on equalizing capabilities, and if capabilities are roughly equalized, then it is still up to individuals to choose how to exercise those capabilities."

"You know what? This still sounds really abstract to me. Is this approach useful? It all depends on what you mean by functionings or capabilities. And is there any way of knowing—or measuring—whether capabilities have been achieved or equalized?"

"You may have a point. There's a big advantage to the welfare approach, or to any approach that tries to equalize *resources*. The advantage is that resources can be defined and measured. Sufficient resources permit a certain level of food or housing or education. It's hard to measure the presence of capabilities. And there's another problem: capabilities can include a huge range of things. So how do we decide which capabilities really matter?"

"So why mention it at all? You brought it up."

"Because this approach has had huge influence. It has influenced thinking on global poverty, and it has influenced United Nations policies on development. It says that you should not just think about how much money or income or material support governments and other agencies need to provide. Instead you need to respect cultural differences. You need to ask what capabilities we want to realize for different people in different situations. And you can start by listing the kinds of capabilities that need support—you can list what you want to put in your capability set.

What capabilities do we want to realize for different people?

"Even more than that, when we think about equality, the *capabilities approach* says we should begin by accepting a universal entitlement to human dignity. What is equality? It is the acceptance of the equal entitlement of all human beings, regardless of their differences, to live a life of dignity. And dignity is not just a vague ideal. It means respect for others as human beings of equal worth. Dignity is respect for humanity as an end in itself and not just as a means to other ends. It is the equality of women and men as citizens, as workers, as human beings. It is health of mind and body."

"So how do you measure that?"

"You don't. Because the highest values can't be measured. They can only be accepted and applied."

17. EQUALITY AND COMMUNITY

"WHAT ARE YOU doing this summer?" asked the professor.

"I'm taking a few weeks off. Planning a canoeing trip. With the same group as last year," replied the plumber.

"Where this time?"

"The Nahanni River in the Northwest Territories. Probably ten days."

"Sounds great. Jeannie going with you?"

"Yeah, if she can get the time off."

"I'm curious. How do you organize it? You'll require all sorts of stuff—canoes, tents, food, pots and pans, portable stoves."

"We figure it out in advance. Everybody contributes something."

"So it's like a group sharing plan. How do you make sure that somebody does not get stuck contributing too much? And others too little?"

"It's not a problem. Because we know who we're going with. We wouldn't include somebody who's likely to be a jerk."

"Why not organize it according to free market principles and the right of private property? If I contribute an expensive item, I assign a price. Others may use the item and offer their items—or a service instead, like cooking—for a price. The exchanges occur across all people and goods. It's efficient. And the result does not depend so much on whether people are jerks or not."

"You airhead! You must be joking! Where did you get that idea from?"

"It's a thought experiment. By a Canadian philosopher named Gerald Cohen. It's called the *camping trip*."

"I don't see the point. It's obvious, isn't it? If you organize it like that, you'd spend a huge amount of time haggling over prices. And the haggling would cause disagreements. So it's not efficient. Besides, it's insulting. It's better to trust people."

"Okay, so let's try to figure out why your system of group sharing works. Is it a socialist system—some type of common ownership?"

"Umm . . . maybe not. Things are shared, but the stuff is still privately owned. On the other hand, our system fits with a principle—something you told me once. From each according to . . .?"

"From each according to her ability, to each according to his need."

"That's it."

"Okay. Set aside whether or not it's socialist. And I'm sure you don't want to hear my lectures on socialism. Is there a principle of equality for your canoeing trip?"

"Sure. Equality of contribution, including effort. The contribution may be big or small, depending on what each person already owns. So the person who already owns a few really good tents contributes them. He doesn't mind if some others contribute in smaller ways. But they'll try hard. They'll probably contribute more food than he does."

"So this equality of contribution doesn't mean that everybody contributes an exactly equal amount of goods or effort."

"No. That wouldn't be possible. Because different members of the group have different advantages. Anyway, there's no way to measure the amounts precisely."

"Does this system allow for different results or different outcomes? It might result in inequalities. That is, some people get fewer benefits than others."

"I'm not sure. . . . Well, everybody gets to participate. Each person shares in the fun and physical challenges—canoeing most of the day, enjoying the beauty around us, enjoying being with friends, sharing the food. But yes, there can be different levels. Like, one person might decide not to go on a hike or see a great waterfall because they're too tired. Or just too lazy."

17. EQUALITY AND COMMUNITY

"So any inequality that occurs is the result of choice. It doesn't result from any lack of access to opportunities or shared goods."

"Ta da! Look what you just did! You just discovered a principle! Is that what this fellow, Cohen, would say?"

"More or less. He's exploring the principles that seem to be at work in the *camping trip*. It's a thought experiment. He knows, of course, that we can't organize the whole of society as though it were a camping trip."

"So what other principles can we get from this?"

"We have to be careful. Any lesson we draw is tentative—*iffy*, you might say. Because of course the people on your canoeing trip are all friends to start with. They're a community—a pre-selected one that exists for a short time only."

"A really cool one, though. I'd call it a co-operative community."

"Right. Because the benefits for each, and for all, depend on active co-operation by each and every member. But what lies behind this desire to co-operate? Are there specific values at work?"

"Sharing. And caring. If there's a value, I suppose, it's caring. We all care—we show interest in the needs and satisfactions of everybody. Right? And that care also applies to ourselves. Each person is caring for herself or himself."

"Exactly! And there's a word for that. Reciprocity. It means an exchange that occurs for the benefit of another. I give—I contribute—and I do so without thinking about how much I'm going to get in return. But guess what? I also benefit because the caring becomes a shared value—a community value. So the caring applies to me too. This caring, this reciprocity, is the core value of community."

"Okay. But that's easy to achieve on a fun canoeing trip among friends."

"And where else?"

"What do you mean, where else? Nowhere else that I can think of. That trip is totally special—it's a kind of utopia."

"Think again. Think outside your box. You're in a box created by a culture where self-interest is taken as the norm. In that box, we give only because we expect to receive. In that box, everything has a price, like in a market. You pay a price; you get something in return. Think outside that box. When do you give, while attaching no price and expecting nothing in return?"

"I gave blood to the Canadian Blood Services a while ago. Got nothing in return, except some satisfaction."

"Good start. Go on."

"When the COVID pandemic was going on, that old friend of mine, who's on a pension, couldn't figure out how to get food and toilet paper. I bought it, paid for it, and got it delivered."

"Don't stop there."

"Don't stop there. Well, you know what—I wasn't doing anything really special during the worst of that pandemic. Lots of people were doing it. Maybe not everybody, but lots of people were. We were supposed to do physical distancing. Even the people who were not at much risk personally. And many of them gave up a lot. The sacrifices were huge. Of course, those who lost their jobs hoped they would get support, or get their jobs back eventually."

"There was, of course, the satisfaction that comes with contributing to the collective benefit."

"And there was a price to be paid if you didn't co-operate. The price was shame—even public shame."

"There you go! Your canoeing trip is not so totally exceptional. There are times—times of catastrophe or collective peril—when reciprocity becomes the norm. A dominant value."

"Aren't you talking about altruism? Totally unselfish concern for others. Acting in ways that involve no self-interest. Does it really exist? And even if it does, surely it's really scarce?"

"Scarce? Really? I think you're still stuck in your box. A box created by simple economistic thinking and market logic. No, altruism, as a belief and behaviour, is central to all the world's major religions. It's not a scarce resource existing in fixed amounts. It's infinitely expandable. The more we use it, the more it grows. It's like virtue. Virtue grows with practice. And as Aristotle said, 'We become just by doing just acts, temperate by doing temperate acts, brave by doing brave acts.'"

"Okay, but what's all this got to do with inequality or equality?"

"We have a principle about when some inequality is acceptable. Inequality is acceptable when it results from personal choices, such as whether or not to work overtime, or whether or not to take unpaid leave, or whether or not to go to college or university when one has the opportunity. Inequality is not acceptable when it results from nothing but bad luck, such as being born into a very poor family. Or from the lack of access to jobs or education or the freedoms that are available to others."

"And equality?"

"Equality is not a singular thing possessed by individuals. It's a relational value. There was no equality for Robinson Crusoe when he was living all by himself on his island. So equality goes with community. Community, with its basic reciprocity, creates equality. Equality is the communal practice of reciprocity."

18. SUCKING UP WEALTH

"DO YOU EVER think about that time?" asked the plumber.

"You mean ... when I came to live with you," replied the professor.

"Yeah."

"There's hardly a day that goes by."

"You were eight, weren't you? They told me I was getting a big sister."

"I kept thinking that my parents would come back.... That they'd suddenly show up and take me home."

"And umm ... my parents ... well ..."

"They had their problems, didn't they? Problems we couldn't begin to understand at the time."

"So we made up things ..."

"Invented ..."

"Imaginary worlds."

"The Planet of the Unicorns ..."

"Where children ruled ..."

"And did whatever they wanted ..."

"Escaping, in a way ..."

"And belonging. And building our futures."

"We were?"

"Yes. Stories make the mind strong. Stories told us what we wanted."

"Is that what we're doing now?"

"Perhaps ... but that was a time of loss and loneliness. And we were trying to survive. Mostly it is loss that teaches us about the worth of things. A guy named Schopenhauer said that."

"Including the worth of equality?"

"Is equality nourished in loss? In a catastrophe? In war, there's a new equality. Equality is proclaimed everywhere—equality of danger, equality of sacrifice, equality of contribution. And during war, a lot of wealth is destroyed."

"In war, we're all in it together. And we heard that so often during the COVID pandemic, didn't we?"

"*Were* we all in it together? What about those who used it as an opportunity to make a profit? Like the couple who bought all the Lysol wipes at a Costco store and then sold them at inflated prices. And made a huge profit."

"That's crazy."

"Why? It's seeing an opportunity, a market demand, and meeting that demand at prices the market will bear."

"It's disgusting! And it should be illegal! Because everybody needs those wipes. And everybody should be able to get them at Costco's prices. Instead two people make a big profit, and only those who can afford the higher prices get the wipes. There must be a word for exploiting people's needs in that way."

"A word for it? Perhaps profiteering? Or parasitism. Which means sucking value from an economy for private gain without contributing anything to production or social good."

"Yeah. Something like that."

"Parasitism is an old word for rent-seeking. That's what economists call it. Here's another example: a really famous one from the First World War. A respected Canadian businessman named Sir Joseph Flavelle was accused of profiteering from war-time demands. Specifically, his meat-packing business had made big profits—perhaps an 80 percent return on capital—from selling bacon and canned meat to governments. There was a huge public outpouring of rage. Flavelle was condemned for profiteering. In fact, the prices for his bacon may not have been inflated. It didn't matter. His profits seemed to violate a principle of equality: equality of sacrifice. In war, all must sacrifice or contribute equally, according to their ability."

"So what happened?"

18. SUCKING UP WEALTH

"The government was forced to regulate the profits of meat-packing businesses. But it was more than that. The Flavelle scandal helped energize a huge public campaign already under way—the campaign for *conscription of wealth*. Conscription of wealth was an equality claim. When men were being conscripted into military service and were asked to sacrifice their lives, then businesses should have been asked to contribute a bigger share of their profits."

"Well, that seems fair."

"But tell me something. The Flavelle scandal, like the story of profiteering during the COVID pandemic, was about a single businessman in a time of crisis. Why would these individual cases provoke outrage? Why do you see it as disgusting when two ordinary people make money by selling Lysol wipes? That kind of profit-making occurs all the time. It's normal behaviour in capitalist markets. Do you condemn the same behaviour when big financial companies and derivative traders and pharmaceutical companies do it all the time?"

"Perhaps it's less obvious when it happens all the time. As you say, it's normal."

"So here's the problem. If this behaviour is morally unacceptable—if it violates a principle of equality—then how do we make it more obvious? How do we make parasitism more apparent even in normal times?"

"I've got no idea. Nobody will see it when it's buried in academic journals and a handful of economics books. What did you call it? Heterodox economics?"

"Did anybody ever try to display the moral problem before a wider public?"

"I don't know. You tell me."

"Never heard of Occupy Wall Street? A big protest movement against the top 1 percent in New York City and many other cities in 2011 and 2012. It was actually an idea from a Canadian magazine called *Adbusters*."

"Of course! I've heard of Occupy Wall Street. What happened to it?"

"It attracted a lot of support in the United States, Canada, and other countries. After a few years, it fizzled out. Why did it fizzle out? The Occupy movement lacked a coherent message and a coherent set of core principles. I think the lesson is this: a principle of social justice—such as the principle of equality—requires two things. First, a clear and coherent understanding of the principle. Second, the principle must be realized in a stable and democratic political movement or formation."

"You're asking for a lot, aren't you? As usual! But I'm hearing something else here. We started off by talking about loss and about catastrophe. Are you saying that these things you want are more likely to happen in a time of crisis—in a war or a pandemic?"

"I think so. More likely to happen when there's a new sense of community—the community of all who are in it together."

"Like brothers and sisters."

"In it together.... Equality: the communal practice of reciprocity."

"But we're not in a war today. And when the vaccines work, the worst of the pandemic will be over."

"What are you saying? We're *not* in the middle of an existential crisis—a threat to existence—like a world war?"

"Okay, wonderwoman. So we're in a crisis. But it's not the same as the Second World War, is it! It's a new and different crisis. So how do we think our way out of it?"

19. RISING ABOVE MYTHS

"SO CLIMATE CHANGE is a threat to our existence?" asked the plumber.

"At worst, yes. We know that already. Climate change is already causing a huge destruction of wealth. And despite the destruction of wealth and capital, climate change may be the cause of inequality in the future. The wealthy can save themselves by moving away from coastlines, by buying up food and other necessities, and by sheltering themselves in air-conditioned mansions. Other people will have a tough time protecting themselves," replied the professor.

"What a pessimist you are!"

"No, I'm not a pessimist. Remember something: when I was a child, real hardship taught me the need for hope and where to find it. Hope requires facing hard truths. Hope also requires learning and thinking—and then acting on one's learning. In learning lies our real wealth. A Greek philosopher named Epictetus made the point long ago. He said, 'Be careful to leave your children well instructed rather than rich, for the hopes of the instructed are better than the wealth of the ignorant.'"

"There you go again! You're full of stuff from Greeks who lived thousands of years ago. But today, we live in a world full of fakes and frauds—fake news, fake knowledge, weird conspiracy theories. There are even people who think that climate change isn't very serious. People who think that COVID is fake news. And people who think that inequality isn't a problem."

"True. So we take action, right? That means spotting fakes and dismantling phony arguments whenever we see them."

"And you're going to tell me that there's a truckload of phony arguments about inequality."

"Such a big truckload that I can show you only a few. So let me give you a few quotes and get your reaction."

1. *"Free markets foster environmental friendliness. If the world is to stave off climate change, then capitalism and the free market will be among its chief allies."*

"What a bonehead idea! Let the fox take care of the chickens."

"I think you're too kind. This fantasy proposes that foxes help build the chicken coops and then manage them.

"It's called free market environmentalism. The theory was that the free market, together with property rights and private legal action against polluters, would effectively reduce pollution and damaging externalities. The theory allowed only a limited role for the state. I use the past tense because the theory is dead, cherished only by a few zealots who believe in ghosts.

"The COVID pandemic taught us a great lesson. Survival demands collective action, and that means direct and forceful action by governments."

2. *"The natural differences in ability, interests, and preferences within any group of people leads to inequalities of outcomes and imperfect living and working conditions that utopias committed to equality of outcome cannot tolerate."*

"But of course there are natural differences among people! Does that lead to extreme inequality?"

"Good answer. But also, who said anything about utopias being committed to equality of outcome? Ever heard of the straw man? This quote is a classic example of the fallacy of the straw man. It's an old debating trick. Present an argument that your debating opponent never offered, and then knock it down. Watch out for this fallacy! It's all over the place. To argue for a reduction of extreme inequality is not to argue for equality of all outcomes—and it's not utopian.

19. RISING ABOVE MYTHS

"And another thing: watch out for the pessimistic idea that human nature offers insurmountable obstacles to social reforms. The argument is that big reforms defy human nature, and so they inevitably fail. There are many ways to answer this objection. One answer is: whatever human nature may be, history tells us that many reforms *do* succeed. Another answer is: human nature and evolutionary endowments are not fixed or constant. They change, and they can *be* changed. Don't just sit there. Evolve!"

> 3. *"Inequality is the mechanism through which the market generates and spreads innovation, which in turn generates opportunities for millions of individuals."*

"I don't know what to think about this one. Innovation is important, isn't it? I suppose it depends on whether inequality leads to innovation."

"Yes. And the origins of innovation have little to do with inequality. But there's another problem. The quote says that the market generates innovation. But that's just plain wrong. Innovation is generated from a complex interaction among many factors—the state, research infrastructures, education systems, and much more. In 2013, the economist Mariana Mazzucato wrote a book called *The Entrepreneurial State*. She debunked the idea that innovation comes from heroic individuals like Steve Jobs or from the Silicon Valley venture capital. Innovation occurs from a chain of investments and risk-taking—a chain that includes state-funded research, applied research, public funding, and more.

"She says, 'The advent of all the important technological changes of the last century—all the high-risk and high-capital investments—came from the State.'"

> 4. *"They are casting their problems at society. And, you know, there's no such thing as society. There are individual men and women and there are families. And no government can do anything except through people, and people must look after themselves first."*

"There's no such thing as society? Really? Who would say that?"
"The British Prime Minister Margaret Thatcher. These words are

often quoted. They are the most extreme expression of neoliberalism from that era, three decades ago."

"Of course people must look after themselves. But let's get real! I sure can't build highways or hospitals or schools by myself."

"You're right. And let me offer an answer to Margaret Thatcher, from the better angels of our nature. From the strong, tough heart of our religious and secular traditions—where today instead of *man*, we would use the more inclusive *one*:"

> No man is an island
> Entire of itself;
> Every man is a piece of the continent,
> A part of the main . . .
> And therefore never send to know for whom the bell tolls;
> It tolls for thee.

5. *"It seems to me, however, that our most fundamental challenge is not the fact that the incomes of Americans are widely unequal. It is, rather, the fact that too many of our people are poor."*

"We talked about this before, didn't we? When we discussed the welfare state."

"Yes. The argument is often rooted in genuine humanitarian concern. But there are too many problems with this idea. For one thing, reducing or eliminating poverty would not get rid of the dangerous overlap between extreme inequality and the unequal distribution of political power. Focus only on poverty, and you are not solving the problem of inequality. You are avoiding the problem."

6. *"A deep and important contribution of the discipline of economics is that greed is neither good nor bad in the abstract."*

"I thought that greed was one of the seven deadly sins."

"It is, in Christian ethics. My comment? The hubris of this quote is breathtaking. Greed is not an abstraction. Greed is behaviour that is undertaken purely for self-interested accumulation, without regard for others. Greed has been condemned in all major religions

19. RISING ABOVE MYTHS

for millennia. Along came some late 20th-century economists, with their selective reading of Adam Smith, who declare that economics can determine for us whether, and when, greed is good or not, through its models and findings. No other scholarly discipline would make such a claim for itself.

"This quote is not an aberration, and much of economics serves as an apology for greed. Little wonder that an abundance of research has shown that studying economics actually makes people think and behave more selfishly."

> 7. *"Philanthropy, which translates as 'love of humanity,' is often presumed to be 'good' by definition. It is also widely understood to be redistributive because it takes money from the wealthy and uses that money to improve conditions for those who are less fortunate."*

"Yeah, I know. Didn't Mark Zuckerberg say that he would give away 99 percent of his Facebook shares?"

"Philanthropy does nothing to solve the problem of inequality. In Canada, all charitable giving was only 0.77 percent of the GDP in 2016—and that includes all donors, not just the wealthy. So it's too small to have redistributive effect. In the United States, there's an inverse relationship between rising inequality and charitable giving by higher-income people; that is, the more inequality, the smaller the donation."

"Yeah, well, I still want to donate to the food bank. Especially in December."

"Me too! But let's not delude ourselves. Giving to food banks doesn't solve the problem of food shortages. Because the basic problem isn't a lack of food. It's poverty."

"But if lots of very rich people gave lots of money, wouldn't that help?"

"No! That would just hide the problem even more. It's called philanthrocapitalism. The philanthropy of the super-rich serves to sanctify wealth. It tries to collapse the distinction between private and public interests, in order to justify increasing concentrations of wealth. It's another way that the super-rich try to capture control of the public sector.

19. RISING ABOVE MYTHS

"Philanthrocapitalism is not just an excuse. It's a huge confidence trick."

"That's just depressing. I thought you were looking for hope!"

"I am. And I am giving you hope."

"You are?"

"Yes. Now you can see, more clearly than before, where the real pessimism lies, naked and exposed to us all. It's the emperor with no clothes. You can see why we must solve collective problems collectively—through the politics of belonging. And through stronger democratic institutions in which all citizens have their say. In that way of seeing, hope flourishes."

20. ALTRUISM AND A GAME

"LET'S PLAY A game, before we have dinner and before you drink too much beer," said the professor.

"I don't like your games. When we were kids, you invented games that you always won," complained the plumber.

"And took a lot of money off you, didn't I?"

"Yeah, though you usually gave it back. Most of it, anyway."

"Don't worry, there's no trick to this game. And it's not really a game because you don't have to do anything. But to play, we have to pretend that we don't know each other at all. We can't even see each other."

"Another veil of ignorance!"

"Exactly. So to start, I'm going to get a large sum of money, on condition that I share it with you. It's totally up to me what portion I give you. You have no say in the division."

"And you don't know it's me who you have to share it with?"

"That's right. If I knew it was you, I'd give you nothing for sure."

"I'm not sure about that. Anyway, what happens next?"

"You have to guess how much I will give you. I could give you any amount—almost nothing, or half of it, or all of it, whatever. Remember, I don't know you."

"And I don't know you?"

"Right. Imagine I'm any person chosen randomly from the population. Also, nobody else knows what decision I'll make."

"Okay. Well, then, it seems like it won't cost you if you keep it all. So I predict you would keep it all."

"Wrong answer! This game is called the dictator game or ultimatum game. The person making the decision is usually called the

20. ALTRUISM AND A GAME

dictator. The game has been played as an experiment many, many times. The average amount of money given is between 20 percent and 30 percent—that's according to the psychologist Dan Ariely, who is an expert on this game. Most dictators give some money, and the result spans across countries, genders, and ages. If the dictator has previously met the other person, or knows that they will meet them at a later time, or knows that the other person is in need, the percentage shared will rise."

"That's wild! But it's just an experiment. What does it tell us?"

"An experiment on real people. That's its advantage. But what does it tell us? There's a huge debate about this question. And the simple dictator game was only the beginning. It's part of a huge field called behavioural economics."

"But isn't it obvious? The experiment tells us that people are not always selfish."

"True. But did behavioural economics kill off homo economicus? Did it kill the model of economic *homo* as a rational maximizer of utility? Many people concluded that it did. On the other hand, many others—including Conservative governments—tried to use this new psychology in manipulating people and in widening the understanding of utility. But there were other problems with homo economicus. He was always a male—a masculinist conception—who crowded out other conceptions of humans as actors. Such as homo reciprocans—humans as co-operative actors. Or homo caritatis—humans as caring actors."

"Now you sound like a feminist. Which I guess you are."

"And we never finished talking about pay equity, did we? Because you fell asleep! Anyway, here's the point. If we want to understand equality, we have to listen to what generations of feminists have been saying. Because for generations, they have been challenging the deep bias of neoclassical economics and its ideas about human motivation—the idea that society consists of self-interested, solo individuals pursuing material gain. Long ago, feminists knew that humans were connected actors, moved by the complex needs of family and community."

"I thought feminism was about equal rights for women."

"It is, but it's much deeper than that. Think of it as a mindset or a value set. In that mindset, equality is universal. It applies to all, or it is meaningless. You can't think about economic inequality unless you also think about gender equality. And racial equality as well—an end to racism."

"My God, are you ever asking for a lot! You're a tyrant, you are. . . . Nothing new about that! You always were really bossy."

"Perhaps that's because you were such a little shit. But really, all I'm saying is this: we've got to accept that human behaviour and motivation include large amounts of empathy and even altruism. Some scholars have suggested that empathy is genetic. Perhaps it's a genetically inherited survival mechanism, as well as a product of social and cultural conditioning."

"Empathy is in our genes? How could you ever prove that?"

"Not easily. But let me ask you something. Human beings are mammals, right? Like some other mammals—the elephants, the great apes, some whales—we have large brains, we are self-conscious, and we communicate. But there's something else—how do we reproduce?"

"Females give birth to babies."

"Yes, but only a very few babies. The babies take a very long time to mature. A newborn baby can't survive on its own. And long-term survival requires maternal self-denial, the support of others, and huge investments in survival skills. Reproduction requires co-operation and self-denial—giving food to another person before taking food for oneself. Remember another thing: for most of our history, we were hunters and foragers. And communities of hunters and foragers had to practice equality and co-operation. They had to share the benefits of luck. Otherwise, they would not survive. Does this mean that empathy is coded into our genes? That's a matter of debate. But we can say that humans have an enormous capacity for empathy and co-operation. That capacity is just as prominent as the capacity for reason."

"All very interesting. But what's all this got to do with inequality?"

"A lot. Because empathy and altruism are crucial to ensuring reciprocity—the communal practice of exchange, of giving and sharing.

20. ALTRUISM AND A GAME

And remember what we said? Equality is the communal practice of reciprocity.

"It turns out that humans have something called inequality or inequity aversion. People will sacrifice benefit for themselves in order to avoid unequal outcomes for others—to avoid what they believe is *unfair*. These findings are really important. They offer a basis for hope. Certainly they make nonsense of the claim that hope for greater equality is merely wishful thinking or utopianism."

"But if this so-called inequality aversion exists, why don't we see more of it? Why doesn't it stop people from accepting the inequality we see today?"

"It does! And this is really important. The inequality we see today is *not* accepted.

"Let that sink in. Step outside the mental box created by much of our media, by neoliberal ideology, by a culture that sanctifies wealth. The inequality we see today is *not* acceptable—and it is *not* accepted.

"This is where research—where social science—is indispensable. It cuts through the myths and fantasies that impair our understanding of the world.

"Take Americans, to start with. Americans from all walks of life say that they accept some degree of inequality. But nothing like the inequality that actually exists. How much should CEOs earn? Americans' *ideal* ratio of CEO pay to employee pay was 7 to 1, according to a survey published in 2014. The *actual* ratio was 354 to 1. The study also revealed that Americans did not have any idea what the ratio actually was. On average, they estimated it to be 30 to 1.

"And Canadians? Canadians know, and say, that inequality has increased significantly in recent decades. And they don't like it. According to a 2017 survey, 82 percent said that they thought income inequality was growing, and 84 percent said that it was a problem. They also believed that the problem could be solved: through a more progressive tax system, a return to pre-Stephen Harper corporate taxes, closure of tax loopholes, and other changes in government policies.

How much will she share with him?

20. ALTRUISM AND A GAME

"Canadians say all this, despite the fact that most underestimate the actual levels of inequality. For wealth (not income) inequality, for instance, a survey in 2014 showed that Canadians thought that the wealthiest fifth held just over half of the wealth in Canada (55.5 percent). In fact, the wealthiest fifth at that time held two-thirds (67.4 percent) of the wealth. And how much *should* the richest 20 people hold? The answer: 30 percent.

"The Occupy movement of a decade ago came out of something real. It mobilized knowledge—and an ideal of equality—that was strong and durable. Here is where the politics of equality began."

21. AN EQUALITY MANIFESTO

"YESTERDAY..." said the professor.

"All my troubles seemed so far away," continued the plumber.

"You may enter."

"My troubles... and we had our share, didn't we? Good thing you made me finish high school."

"You just needed a push."

"A kick in the ass, you mean. Remember that parent-teacher meeting? When you showed up? The look on that teacher's face! You were only nineteen."

"In loco parentis. He soon got the picture."

"Well, yeah! You let him know who was in charge. And then you threatened to stop cooking for me."

"And I meant it."

"Yeah, I know. And the threat worked. I had a huge appetite. Anyway, you'd already been looking after me. And we moved out and lived in that dingy apartment."

"Didn't have much of a choice, did I? I couldn't go off on my own and leave you to screw up."

"Do you scare your students like that?"

"Don't need to. I make them laugh. Then I give them a problem they don't know how to solve. Drives them nuts. But they learn."

"No wonder you won that teaching award."

"Students energize me."

"As if you need any more energy."

"To teach is to learn."

"Really? But you seem to know everything already."

21. AN EQUALITY MANIFESTO

"Nope. I work with ideas. Ideas demand communication. We do not know, in any meaningful way, until we talk to each other. The idea does not exist until it is put into words."

"You're the one with the words. So that means you produce the ideas."

"I don't know about that! I would not produce my ideas through words, without the audience, real or imagined, that hears me. So by listening, and thinking and reacting, my students participate in the creation of the ideas. So do you. After all, who's been leading this conversation?"

"You have! You tell me lots of stuff in simple words, and I listen."

"Hang on! *You* have been steering this conversation right from the beginning. You first raised the issue of what *stuff* should be distributed in an equitable way. You said you wanted people, not numbers. You said that principles must precede policies. You grounded those principles in an idealist philosophy. You were the one who expressed reservations about John Rawls. You know a lot about inequality. And you, little buddy, are a homo reciprocans."

"Oh yeah? Nobody's ever called me that before!"

"So don't tell me that I'm the boss here. There's an equality in our community of two. Even though I could beat you up when we were kids ... accepting the presence of the other as necessary, essential to one's own life and activity, is important."

"There you go! Another definition of equality. Sounds cool to me."

"And you're not totally stunned, are you? You offer ideas too. So what does equality mean to you?"

"Didn't the Americans say something about it? We hold these truths to be self-evident, that all men are created equal. And the French too: liberty, equality, fraternity!"

"Eighteenth-century statements, coming from the American and French revolutions. But they can't serve as well today, can they? Is there a problem with the idea that all men are created equal?"

"Well, yeah! What about women? Or what about African-Americans? What about Indigenous peoples?"

"So give me a principle."

"Equality—it means treating people the same regardless of gender or race or any other aspect of their identity."

"The same? But perhaps some people have different starting points and different needs. What about racialized people who start from a position of disadvantage? Would you treat them the same as white people?"

"I guess not. So give people the same respect because they are human, while taking into account their different needs."

"So you're saying equality is respect for the equal worth and equal dignity of all human beings. And what about different needs? You could add *equality of consideration*. That is: equality is respect for the worth and dignity of all, while giving consideration to the different needs and interests of those affected by one's actions."

"Okay, that's all fine. But where have we got to? We're supposed to be talking about inequality, especially extreme inequality."

"So we've got to somehow connect our principles to distribution. And that's not easy, for sure. And you can't separate other inequalities—those relating to gender and race, for instance. Your equality has to be inclusive; otherwise it's worth nothing. Reciprocity accepts and respects differences; otherwise it does not exist."

"That's a huge problem, isn't it? What happens during a pandemic or when climate change hits us even more? How can you distribute risks and costs and losses? I don't see how that can be done."

"One step at a time. There are burdens. Burdens when we strive for mitigation or prevention—trying to keep the global temperature increase to 2 degrees or less. And there are costs involved in adapting to the effects of climate change. What is a fair distribution of these burdens?"

"It's only fair that those who create most of the greenhouse gas emissions pay for the cost."

"Only those who create such emissions? That's called the Polluter Pays principle. But what about those who benefit from carbon-based fuels? And we all benefit, to some degree, don't we? We're all complicit. So how do you determine who the polluter *is*?"

21. AN EQUALITY MANIFESTO

"Okay, but there are ways of measuring emissions. Everybody should have a carbon footprint calculator, including companies. And everybody should be allowed a minimum level of emissions—the same for all. It's fair, and it's equal too!"

"So what do you do about those who exceed their permitted level? Some companies will have to exceed their permitted level because of the nature of their businesses. Perhaps some should be allowed higher levels. Or should they be allowed to trade permits with others with lower levels?"

"That's easy. Trading away your carbon emissions defeats the purpose. Those who exceed their level must pay. Not the same amounts but rather according to their ability to pay. Equality of contribution: contributing according to your ability to contribute. So the Equal Emissions policy becomes part of the tax system."

"So let me try to understand this. Rich people have big carbon footprints. They fly to meetings in their own jets. Large numbers fly to Davos every year. They rent big gasoline-powered cars to get to the meetings and so on. We know that they don't have to do any of this. The COVID pandemic proves that important meetings—even conferences—can be held virtually, through increasingly sophisticated meeting apps. So the meetings aren't a necessity. They're a choice. Perhaps even a luxury."

"Yeah, but they can still do it if they want to."

"Right. So freedom of choice is not denied. But equality takes priority over a costly liberty. They can still do it. But equality of contribution requires that they pay. Their excess emissions become a tax."

"So there's still something left of the Polluter Pays principle. But there's also the Ability to Pay principle. Because the tax system is progressive, right? Meaning that those with big incomes pay a much higher percentage of their incomes in tax."

"Yes, of course. But that means *real* progressive taxation across the whole tax system. It means returning top marginal tax rates to where they were 50 years ago. And it means a war on tax evasion. Equality means Fair Taxation. It also means equality of treatment

121

of money. If you tax income from employment, you should also tax income from capital and from inheritance."

"Do you think this is feasible? Isn't it utopian?"

"Would you prefer extinction? Or a planet with average temperatures 3 or 4 degrees higher than where they are now?"

"But you're infringing on a basic right. The right of ownership and use of private property."

"Haven't you forgotten something?"

"Of course, yes, I remember. The right of ownership is not unlimited."

"Exactly. It's subordinate. We hold the first principle of justice to be equality, which means respect for the equal worth and dignity of all. This equality is not individual but communal, so then follows reciprocity: the necessary actions of giving and taking for mutual support. Then follows the right of ownership: private ownership is consistent with equality and liberty, but only when it is a limited and contingent right. You cannot build a skyscraper on your land because to do so would infringe upon the rights of others to the use and enjoyment of their land.

"So is it a just exercise of ownership when a rich entrepreneur builds rockets to fly tourists to the moon? Regardless of the huge costs to humanity in terms of emissions and in terms of capital diverted into a self-indulgent luxury? Is it a just exercise of ownership for people to purchase and own essential medical equipment during a global pandemic and then sell the equipment at inflated prices?"

"No."

"Why not?"

"Because such gross self-indulgence is no longer merely selfish. It has become an unnecessary harm to others."

"If you do not change direction, you may end up where you are heading."

"We're all in it together. What we heard so often during the pandemic."

21. AN EQUALITY MANIFESTO

"Reciprocity: the core value of community. But how do we know if we're getting there?"

"We can't know for sure. There's no exact measure. But we can do a lot better than Occupy Wall Street and its slogan: the 1 percent versus the 99 percent. There are better ways of thinking and speaking. Companies can track, and limit, the ratio between CEO compensation and employee wages—and shareholders and regulators can insist that they do so. And we can do something like this for all incomes. When it comes to income, where does yours rank among incomes in your community? The median after-tax income in Canada is around $62,000. Put yourself on the income scale. I know you don't like numbers, but do this for me. Here's a calculator. Divide your after-tax income by $62,000 and tell me what you get."

"I get 0.76."

"Okay. For myself, I get 1.4. Compare that to the two people I told you about when we started discussing all this. Emily is at 0.4. Michael is somewhere between 11 and 15, though we can't know for sure. And if we were using his total before-tax remuneration, his number would be 177."

"That's way too much."

"*What* is too much? I'll tell you. The spread—the gap—is too big. We will know that we are succeeding when that income spread shrinks. There's no such thing as perfect equality. But reciprocity means a closing of distance—a coming together. It means bringing both Emily and Michael back from their estrangement and back into the community of which they should both be members.

"That closing of distance is a moral norm in the pandemic, even as we practice what we call physical distancing. We all search for connection in distressing times. And the moral norm has been embedded in law, however imperfectly, in many ways, at many times. As it often was in war. The moral norm—the pre-eminent demand for equality—frames the rights and duties of all: individuals, groups, organizations, corporations."

"All in it together."

21. AN EQUALITY MANIFESTO

"So now we can go back to the policy solutions we talked about before. Armed with principles of justice."

"The state has the power and the duty to extinguish the distances that inequality imposes."

"Reciprocity . . . it comes out of necessity. As you and I learned long ago. In our *yesterday*."

"So we build the politics of survival on the politics of equality."

"Shame falls upon those who go it alone."

"Waste, in a time of universal need, is unacceptable harm."

"Wealth is desanctified. And when that happens, we bury the idea that inequality is inevitable."

"The worth and dignity of others is your own."

"And right now, my worth and my dignity require another beer."

"Your need is my wish, brother."

"And we've done okay, haven't we? Thanks to your beer and our cooking. And to the food of thought."

"The food that gives us power."

"Exactly! Good night, cousin Angus."

"Good night, Rachel. Get out there and solve some big problems."

"Only with you."

SOURCES

I HAVE CHOSEN not to use footnotes but to list some references for each chapter here. These references are intentionally very selective. For almost every issue raised in this book, there is a substantial body of writing.

1. LET'S TALK

Readers of Plato will see that my opening has echoes of the beginning of *The Symposium*. In this book, the participants in the opening dialogue were not all sober when they arrived, but they agreed not to drink too much during the dinner. They dismissed the flute player.

Gini coefficient: the Gini coefficient is equal to the area marked A divided by the sum of the areas marked A and B—that is, Gini = $A/(A + B)$. The result always falls between 0 and 1. This measure was developed by the Italian Corrado Gini and first published in 1912. It has advantages: it is simple to understand, and it allows for easy comparisons across time and among different geopolitical units. Although used very frequently, it has disadvantages. For instance, the coefficient gives more weight to the middle of the distribution and little weight to changes at the end of the distribution.

SOURCES

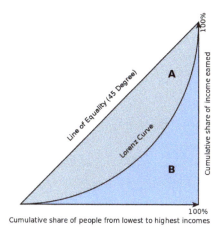

Gini coefficient

"It depends on what stuff we're talking about": our astute plumber raises a deep concern in the field of distributive justice. This book starts but does not conclude the discussion of the distribuendum—that which is to be distributed. The emphasis here is very much on income (and, to a lesser extent, wealth).

2. EMILY, A WORKER

According to *Indeed: Job Search Canada*, the average hourly wage for a warehouse worker in Ontario in February 2020 was $16.22. As an inexperienced employee, Emily's wage is closer to an entry-level salary of around $14.00 (the minimum wage in Ontario). In early 2020, rental costs in London, Ontario, for a two-bedroom apartment were between $900 and $1,500. *Numbeo* estimated the single-person living costs (without rent) for London, Ontario, to be $997.26 (April 2020): https://www.numbeo.com/cost-of-living/in/London-Canada. *Expatistan* estimated the single-person monthly living costs for London, Ontario, to be $2,149 (April 2020): https://www.expatistan.com/cost-of-living/london-ontario.

Poverty organizations offer an abundance of information on low-income families. The official source is Statistics Canada's Dimensions of Poverty Hub: https://www.statcan.gc.ca/eng/topics-start/poverty.

Healthy Smiles Ontario is a government-funded program providing dental services for children in low-income families. If Emily persisted, she would probably qualify.

3. MICHAEL, A CEO

Canada's current (as of August 2020) minister of finance wrote about the lives of the super-rich: Chrystia Freeland, *Plutocrats: The Rise of the New Global Super-Rich and the Fall of Everyone Else* (Toronto: Doubleday Canada, 2012).

On executive compensation: David Macdonald, *Climbing Up and Kicking Down: Executive Pay in Canada* (Canadian Centre for Policy Alternatives, 2018); Canadian Centre for Policy Alternatives, "High CEO pay shatters previous records, now 227 times more than average worker pay," news release (2 January 2020), https://www.policyalternatives.ca/newsroom/news-releases/high-ceo-pay-shatters-previous-records-now-227-times-more-average-worker-pay.

There is plenty of information online about CEO work routines. See, for instance: "9 CEOs Share their Morning Routines," *Fortune* (5 September 2017), https://fortune.com/2017/09/05/ceo-morning-routine-success/.

A study of CEO time: Michael E. Porter and Nitin Nohria, "How CEOs manage time," *Gordon Institute of Business Science Blog* (1 July 2018), https://gibsic.wordpress.com/2018/07/01/how-ceos-manage-time-by-michael-e-porter-and-nitin-nohria/.

On pay increases: "Canadian CEOs make 200 times more than average worker," *phys.org* (3 January 2020), https://phys.org/news/2020-01-canadian-ceos-average-worker.html.

On change over time: Lars Osberg, *The Age of Increasing Inequality: The Astonishing Rise of Canada's 1%* (Toronto: James Lorimer & Company, 2018), 175–81.

The story of Sandor: John Marlyn, *Under the Ribs of Death* (London: A. Barker, 1957).

4. THE RISE OF INEQUALITY

Graphing the top 1 percent in Canada, the graph reproduced here is from: Michael R. Veall, "Income inequality has risen in Canada, but not in a straight line," *Policy Options* (9 March 2020). I am grateful to Michael for granting me permission to use this graph. See also Michael R. Veall, "Top Income Shares in Canada: Recent Trends and Policy Implications," *Canadian Journal of Economics* 45 no. 4 (November 2012), 1247–72.

On changes in distributions over time in Canada: Emmanuel Saez and Michael R. Veall, "The Evolution of High Incomes in North America: Lessons from Canadian Evidence," *American Economic Review* 95, no. 3 (2005); David A. Green, W. Craig Riddell and France St-Hilaire, *Income Inequality: The Canadian Story* (Montreal: Institute for Research on Public Policy, 2016); Lars Osberg, *The Age of Increasing Inequality: The Astonishing Rise of Canada's 1%* (Toronto: James Lorimer & Company, 2018); David A. Green and Jonathan R. Kesselman, eds., *Dimensions of Inequality in Canada* (Vancouver: UBC Press, 2006); Keith Banting and John Myles, eds., *Inequality and the Fading of Redistributive Politics* (Vancouver: UBC Press, 2013).

The body of published writings outside of Canada is immense, but discussion usually begins with two books by Thomas Piketty: *Capital in the Twenty-First Century*, trans. Arthur Goldhammer (Cambridge, MA: Harvard University Press, 2014); *Capital and Ideology*, trans. Arthur Goldhammer (Cambridge, MA: Harvard University Press, 2020). He did write a shorter book! Thomas Piketty, *The Economics of Inequality*, trans. Arthur Goldhammer (Cambridge, MA: Harvard University Press, 2015).

On neoliberalism see: David Harvey, *A Brief History of Neoliberalism* (Oxford: Oxford University Press, 2005); Philip Mirowski and Dieter Plehwe, eds., *The Road from Mont Pèlerin: The Making of the Neoliberal Thought Collective* (Cambridge, MA: Harvard University Press, 2009).

On homo economicus: Peter Fleming, *The Death of Homo Economicus: Work, Debt and the Myth of Endless Accumulation* (London: Pluto Press, 2017).

On the "fading" of redistributive politics in Canada before 2013: Keith Banting and John Myles, eds., *Inequality and the Fading of Redistributive Politics* (Vancouver: UBC Press, 2013).

On the need to reject economic fatalism and the widespread assumption that inequality is inevitable, see: Eli Cook, "Naturalizing Inequality: The Problem of Economic Fatalism in the Age of Piketty," *Capitalism: A Journal of History and Economics* 1, no. 2 (Spring 2020), 338–78.

5. DOES IT MATTER?

The relationships among inequality, culture, status, and envy are a very old subject. Jean-Jacques Rousseau thought that envy pervades commercial societies because envy increases with wealth and inequality. Rousseau wrote the seminal modern work: *Discourse on the Origin and Basis of Inequality Among Men* (1755). Although not explicitly about inequality, Thorstein Veblen's best-known book is a seminal American study of wealth and culture, and it is still worth reading today: *The Theory of the Leisure Class: An Economic Study of Institutions* (New York: Macmillan, 1902). Even a quick online search will uncover varying interpretations and ideological differences over inequality and envy. There are many American studies of social anger, and psychologists have also offered input.

On social epidemiology and inequality: Richard Wilkinson and Kate Pickett, *The Spirit Level: Why Greater Equality Makes Societies Stronger* (New York: Bloomsbury Press, 2009).

On climate change and inequality, much of the discussion is about global inequality. See, for instance, "Climate change and inequalities in the Anthropocene," chapter 5 in *United Nations Human Development Report 2019*, http://hdr.undp.org/sites/default/files/hdr2019.pdf; Noah S. Diffenbaugh and Marshall Burke, "Global warming has increased global economic inequality," *Proceedings of the National Academy of Sciences of the United States of America* 116, no. 20 (May 2019), 9803–13.

On climate change denial and attitudes toward redistribution: Angelo Panno, Giuseppe Carrus and Luigi Leone, "Attitudes towards Trump Policies and Climate Change: The Key Roles of Aversion to Wealth Redistribution and Political Interest," *Journal of Social Issues* 75, no. 1 (March 2019), 153–68.

On corporate tax evasion: "Tax Gap and Compliance Results for the Federal Corporate Income Tax System," *Canada Revenue Agency* (2014), https://www.canada.ca/en/revenue-agency/corporate/about-canada-revenue-agency-cra/tax-canada-a-conceptual-study/taxgap-compliance-results.html?utm_source=news&utm_medium=news&utm_campaign=Tax_Gap_2019&utm_content=2019-06-17_0031#toc25.

See also: Emmanuel Saez and Gabriel Zucman, *The Triumph of Injustice: How the Rich Dodge Taxes and How to Make Them Pay* (New York: W.W. Norton, 2019).

On the number of lobbying contacts with the federal government: Bill Carroll, Nicolas Graham and David Chen, "Fossil Fuel Lobby Influence Runs Deep," *Monitor* (Canadian Centre for Policy Alternatives, January/February 2020), 21.

On the super-rich and democracy in America, there are many studies, including: Nancy MacLean, *Democracy in Chains: The Deep History of the Radical Right's Stealth Plan for America* (New York: Viking, 2017).

On inequality and voter turnout: Matthew Polacko, "Income Inequality and Voter Turnout in Canada 1988–2011" (2017), http://www.progressive-economics.ca/wp-content/uploads/2007/06/Matthew-Polacko-PEF-Essay-revised.pdf; Antonio M. Jaime-Castillo, "Economic Inequality and Electoral Participation: A Cross-Country Evaluation" (paper for the Comparative Study of the Electoral Systems Conference, Toronto, 2009).

On inequality and right-wing populism: Austin Botelho, "The Short End of the Stick: Income Inequality and Populist Sentiment in Europe," *Issues in Political Economy* 28, no. 1 (2019), 39–78; Sarah Jay et al., "Economic inequality and the rise of far-right populism: a social psychological analysis," *Journal of Community & Applied Sociology* 29, no. 5 (September/October 2019), 418–28.

6. DESPAIR AND DEATH

"Growth has slowed down or stalled." The "end of growth" has generated a large literature, but a seminal work of history is: Robert J. Gordon, *The Rise and Fall of American Growth* (Princeton: Princeton University Press, 2017). See also: Danny Dorling, *Slowdown: The End of the Great Acceleration—and Why It's Good for the Planet, the Economy, and Our Lives* (New Haven: Yale University Press, 2020).

On "deaths of despair" in the United States: Anne Case and Angus Deaton, *Deaths of Despair and the Future of Capitalism* (Princeton: Princeton University Press, 2020).

On life expectancy of homeless people: "Deaths of People Experiencing Homelessness: 2017 Review," *Toronto Board of Health*

(11 April 2018), https://www.toronto.ca/legdocs/mmis/2018/hl/bgrd/backgroundfile-113980.pdf.

See also: "Dying on the Streets: Homeless Deaths in British Columbia" (2014), https://d3n8a8pro7vhmx.cloudfront.net/megaphone/pages/7/attachments/original/1415231881/Dying_on_the_Streets_-_Homeless_Deaths_in_British_Columbia.pdf?1415231881.

On inequality, health, and mortality in Canada: Ibrahima Bocoum et al., "Effect of Income Inequality on Health in Quebec: New Insights from Panel Data," *Sustainability* 11, no. 20 (October 2019); Michael Tjepkema, Russell Wilkins and Andrea Long, "Cause-specific mortality by income adequacy in Canada: A 16-year follow-up study," *Statistics Canada Health Reports* 24, no. 7 (2014); Dennis Raphael and Toba Bryant, "Income Inequality is killing thousands of Canadians every year," *Toronto Star* (23 November 2014).

7. WE NEED INCENTIVES

On CEO motivations: "What Motivates a CEO?" *Incentive & Motivation*, http://incentiveandmotivation.com/what-motivates-a-ceo/; Xavier Baeten, "CEO Motives," *Vlerick Business School* (2016), https://www.vlerick.com/~/media/corporate-marketing/our-expertise/pdf/CEOMotivespdf.pdf; Jean McGuire et al., "CEO Incentives and Corporate Social Performance," *Journal of Business Ethics* 45 (July 2003), 341–59; Dan Cable and Freek Vermeulen, "Stop Paying Executives for Performance," *Harvard Business Review* (23 February 2016); Steven Clifford, *The CEO Pay Machine: How It Trashes America and How to Stop It* (New York: Blue Rider Press, 2017). See also: John E. Roemer, "Ideology, Social Ethos, and the Financial Crisis," *Journal of Ethics* 16, no. 3 (2012), 273–303.

There is also evidence of a "discouragement effect" in the presence of inequality, leading to lower work effort: Hyejin Ku and Timothy C. Salmon, "The Incentive Effects of Inequality: An Experimental

Investigation," *Southern Economic Journal* 79, no. 1 (2012), 46–70. Charles S. Jacobs argues the case for intrinsic motivation rather than monetary rewards: *Management Rewired* (New York: Penguin, 2009).

On Japanese CEO pay: "Japanese CEO pay roughly 10% that of US counterparts," *Nikkei Asian Review* (13 November 2015); Luyao Pan and Xianming Zhou, "CEO Compensation in Japan: Why So Different from the United States?" *Journal of Financial and Quantitative Analysis* 53, no. 5 (October 2018), 2261–92.

8. FIXING THE PROBLEM

Proposed reforms are to be found in: Anthony B. Atkinson, *Inequality — What Can Be Done?* (Cambridge, MA: Harvard University Press, 2015); Joseph E. Stiglitz, *The Price of Inequality* (New York: W.W. Norton, 2012); Joseph E. Stiglitz, *The Great Divide: Unequal Societies and What We Can Do About Them* (New York: W.W. Norton, 2015); Thomas Piketty, *Capital in the Twenty-First Century*, trans. Arthur Goldhammer (Cambridge, MA: Harvard University Press, 2014); *Capital and Ideology*, trans. Arthur Goldhammer (Cambridge, MA: Harvard University Press, 2020).

For global inequality, one starts with Branko Milanovic: *Global Inequality: A New Approach for the Age of Globalization* (Cambridge, MA: Harvard University Press, 2016); *The Haves and the Have-Nots: A Brief and Idiosyncratic History of Global Inequality* (New York: Basic Books, 2010).

For Canada especially, Lars Osberg, *The Age of Increasing Inequality: The Astonishing Rise of Canada's 1%* (Toronto: James Lorimer & Company, 2018); David A. Green, W. Craig Riddell and France St-Hilaire, *Income Inequality: The Canadian Story* (Montreal: Institute for Research on Public Policy, 2016).

There is a substantial literature on a guaranteed or universal basic income. See, for instance: Philippe Van Parijs and Yannick

Vanderborght, *Basic Income: A Radical Proposal for a Free Society and a Sane Economy* (Cambridge, MA: Harvard University Press, 2017); Rutger Bregman, *Utopia for Realists* (New York: Little, Brown and Company, 2014).

"The pen is mightier" and "the enchanter's wand": words borrowed from Edward Bulwer-Lytton's play *Richelieu: Or the Conspiracy* (1839).

9. WHAT WE VALUE

The play by Oscar Wilde is *Lady Windermere's Fan* (1892).

The economics textbook the plumber looked at was: Steven A. Greenlaw and David Shapiro, *Principles of Economics 2e* (OpenStax, 2018), https://opentextbc.ca/principlesofeconomics2eopenstax/.

The world of economics textbooks is changing and unorthodox approaches have appeared. See, for instance: Louis-Philippe Rochon and Sergio Rossi, *An Introduction to Macroeconomics: A Heterodox Approach to Economic Analysis* (Edward Elgar Publishing, 2016).

John Clark Murray (1836–1917) was a Scottish-born idealist philosopher who worked at Queen's University and McGill University. The quotations are from: John Clark Murray, *The Industrial Kingdom of God* (Ottawa: University of Ottawa Press, 1981), 13, 137–8. Although not published until 1981, this book was written in the 1880s.

"The inevitable outcomes of economic laws": Lars Osberg says, "In many university departments of economics, the steady-state properties of imaginary worlds in which, by assumption, poverty and inequality cannot exist at all because these theoretical worlds are entirely inhabited by identical 'representative agents' have been taught to gullible undergraduates since the late 1970s." Osberg, *The Age of Increasing Inequality*, 74.

10. EQUAL OPPORTUNITY

Anatole France's ironic saying is in his story, *Le Lys rouge* [*The Red Lily*] (1894).

Amartya Sen's oft-cited lecture is: "Equality of What?" (Tanner Lecture on Human Values, Stanford University, Stanford, CA, May 22, 1979), http://www.ophi.org.uk/wp-content/uploads/Sen-1979_Equality-of-What.pdf.

See also: Chapter 1 in Sen's *Inequality Reexamined* (Cambridge, MA: Harvard University Press, 1992).

On social mobility: Marie Connolly, Miles Corak and Catherine Haeck, "Intergenerational Mobility between and within Canada and the United States" (National Bureau of Economic Research Working Paper no. 25735, April 2019); Miles Corak, "Income Inequality, Equality of Opportunity, and Intergenerational Mobility," *Journal of Economic Perspectives* 27, no. 3 (Summer 2013), 79–102.

Education relates to human capital, which has generated a substantial literature. For Canada: Kelly Foley and David A. Green, "Why More Education Will Not Solve Rising Inequality (and May Make It Worse)," in *Income Inequality: The Canadian Story* (Montreal: Institute for Research on Public Policy, 2016), 347–97. See also: Lars Osberg, "Chapter 7: Outcomes and Opportunities," in *The Age of Increasing Inequality* (Toronto: James Lorimer & Company, 2018).

T. Phillips Thompson (1843–1933) was an English-born journalist and author and one of the main representatives of first-wave socialist thought in Canada. His sarcastic reference to equal opportunity and prisoners in the Sing Sing prison is in his book (note: published in the United States) *The Politics of Labor* (New York: Belford, Clarke & Co., 1887), reprinted by the University of Toronto Press in 1975.

"Do we deserve our talents?": Michael J. Sandel, *The Tyranny of Merit: What's Become of the Common Good?* (New York: Farrar, Straus and Giroux, 2020), 122–4.

11. REWARDS AND MERIT

A good summary of this field is: Jeffrey Moriarty, "Desert-Based Justice," in *The Oxford Handbook of Distributive Justice* (Oxford: Oxford University Press, 2018), 152–73. Moriarty discusses the connections to "luck egalitarianism"—a subject not explored in the present book.

See also: Richard Arneson, "Dworkin and Luck Egalitarianism: A Comparison," in *The Oxford Handbook of Distributive Justice* (Oxford: Oxford University Press, 2018), 41–64.

Aristotle on merit: *Nicomachean Ethics*, Book V, Chapter 3.

Amartya Sen, "Merit and Justice," in *Meritocracy and economic inequality* (Princeton: Princeton University Press, 2000).

Michael Young, *The Rise of the Meritocracy, 1870–2033: An Essay on Education and Equality* (London: Random House, 1958).

See also: Kwame Anthony Appiah, "The Red Baron," *New York Review of Books* (11 October 2018).

Daniel Markovits, *The Meritocracy Trap: How America's Foundational Myth Feeds Inequality, Dismantles the Middle Class, and Devours the Elite* (New York: Penguin, 2019).

Michael J. Sandel, *The Tyranny of Merit: What's Become of the Common Good?* (New York: Farrar, Straus and Giroux, 2020).

Thomas Piketty discusses meritocratic elitism in France in *Capital and Ideology* on pages 754–9.

12. PAY EQUITY

Alcibiades: in Plato's *The Symposium*, Alcibiades showed up in a drunken state and proceeded to offer frank and sometimes rude comments on the character of Socrates.

The relationship between gender, work force participation, and inequality is immensely complex. See, for instance: Lars Osberg's *The Age of Increasing Inequality: The Astonishing Rise of Canada's 1%* on pages 148–53. There is a vast body of literature on gender and economic inequalities between men and women. Nevertheless, it has to be said that the relationship between gender, women's labour force participation, and society-wide vertical income distributions remains an understudied topic. An example is Ursina Kuhn and Laura Ravazzini, "The Impact of Female Labour Force Participation on Household Income Inequality in Switzerland," *Swiss Journal of Sociology* 43, no. 1 (February 2017), 115–35.

The place of gender in the field of distributive justice is a separate but related subject. A summary is Anca Gheaus, "Gender," in *The Oxford Handbook of Distributive Justice* (Oxford: Oxford University Press, 2018), 389–414.

13. WHAT IS PROPERTY?

Locke and disproportionate possession: "Chapter V: Of Property," in *Second Treatise of Civil Government* (1690).

The history of property is an old subject. See, for instance, Christopher Pierson's three volume series: *Just Property* (Oxford: Oxford University Press); Ellen Meiksins Wood, *Liberty and Property: A Social History of Western Political Thought from the Renaissance to Enlightenment* (London: Verso, 2012); Peter Garnsey, *Thinking about Property: From Antiquity to the Age of Revolution* (Cambridge: Cambridge University Press, 2007).

Bir Tawil: an uninhabited and unclaimed territory on the border between Egypt and Sudan.

Thomas Aquinas (1225–1274): there is a substantial literature about his teachings on property and wealth. A useful summary is: Rachael Walsh, "Property, Human Flourishing and St. Thomas Aquinas," *Canadian Journal of Law and Jurisprudence* 31, no. 1 (February 2018), 197–222.

The related subjects of wealth, property, and social justice remain important preoccupations for Catholic thinkers. Property, moral responsibility, and justice were also important topics for Protestant thinkers and especially for "new liberal" thinkers in the late 19th- and early 20th-centuries. In the 1880s, for instance, the aforementioned idealist philosopher John Clark Murray asked the question, "Does a man, in either law or justice, have the right to do with his property as he pleases?" His answer was an emphatic negative in *The Industrial Kingdom of God* on page 118.

"There's no conversation more boring": a saying attributed to Michel de Montaigne. The nearest equivalent appears to be, "L'accord est une chose tout à fait ennuyeuse dans la conversation": Montaigne, "Chapter VIII," in *Les Essais*, ed. André Lanly (Gallimard, 2009), 1118.

14. WHAT IS FAIR?

John Rawls (1921–2002), American moral and political philosopher. His thinking on justice as fairness had its first full-length expression in: *A Theory of Justice* (Cambridge, MA: Harvard University Press, 1971). See also his books: *Political Liberalism* (New York: Columbia University Press, 1993); *Justice as Fairness: A Restatement* (Cambridge, MA: Harvard University Press, 2001).

For the argument that Rawls's theory of justice was strongly egalitarian, see, for instance: Ian Hunt, "How Egalitarian is Rawls's Theory of Justice?" *Philosophical Papers* 39, no. 2 (July 2010), 155–81.

For a more recent discussion, see: Samuel Freeman, "Rawls on Distributive Justice and the Difference Principle," in *The Oxford Handbook of Distributive Justice* (Oxford: Oxford University Press, 2018), 13–40.

Lane Kenworthy presents hypothetical income data in his book: *Jobs With Equality* (Oxford: Oxford University Press, 2008), 22; see also: Carsten Jensen and Kees van Kersbergen, *The Politics of Inequality* (London: Palgrave, 2017), 13.

15. WELFARE STATES

The story of the clogged clay pipe is adapted from: Briana Jones, "5 Toughest Pipe Bursting Jobs," *Cleaner* (4 December 2014), https://www.cleaner.com/online_exclusives/2014/12/5_toughest_pipe_bursting_jobs.

"Ill fares the land" from the poem: *The Deserted Village* (1770) by Oliver Goldsmith. John Ruskin used the word "illth" as the opposite of wealth in his 1879 book, *Unto This Last*. "Illfare" is an old word—the *Oxford English Dictionary* cites a use of the word in 1474 and "ill fare" even before then.

In the UK, welfare economics emerged from a "new liberal" political economy; its first major exponents were Arthur Pigou and Hugh Dalton. There are many books about the origins of the welfare state. Its limited conceptualization in Canada is discussed in Chapter 9 and Chapter 10 of: Eric W. Sager, *Inequality in Canada: The History and Politics of an Idea* (Montreal: McGill-Queen's University Press, 2020).

On taxes, transfers and inequality in Canada, see: Andrew Heisz and Brian Murphy, "The Role of Taxes and Transfers in Reducing

Income Inequality," in *Income Inequality: The Canadian Story* (Montreal: Institute for Research on Public Policy, 2016), 435–78.

On the weakening of social security, see: Lars Osberg, *Canada's Declining Social Safety Net* (Canadian Centre for Policy Alternatives, 2009); "Canada's Social Spending Is Still Among the Lowest in the Industrialized World," *PressProgress* (16 March 2019), https://pressprogress.ca/canadas-social-spending-is-still-among-the-lowest-in-the-industrialized-world/.

According to the latest OECD data, Canada is 24th among OECD countries in public social spending as a percent of the GDP: "OECD Social Expenditure Database" (2020), https://www.oecd.org/social/expenditure.htm. Canada is at 17.3 percent. France, Belgium, Denmark, Austria, Finland, Italy, Sweden, Germany, and Norway are all at 25 percent or more.

16. EQUALITY AND CAPABILITY

Aristotle on friendship: *Nicomachean Ethics*, Book VIII, Chapter 1.

Capabilities approach: there is a substantial literature. Amartya Sen: *Inequality Reexamined* (Oxford: Clarendon Press, 1992); "Capability and Well-Being," in *The Quality of Life* (Oxford: Clarendon Press, 1993), 30–53; *The Idea of Justice* (London: Allen Lane, 2009).

Martha Nussbaum: "Human Dignity and Political Entitlements," in *Human Dignity and Bioethics* (New York: Nova Science Publishers, 2008), 245–64; *Women and Human Development: The Capabilities Approach* (Cambridge: Cambridge University Press, 2000); *Creating Capabilities: The Human Development Approach* (Cambridge, MA: Harvard University Press, 2011). Nussbaum acknowledges the complexity of dignity as a concept. Among others: Vasil Gluchman, "Human Dignity as the Essence of Nussbaum's Ethics of Human Development," *Philosophia* 47 (2019), 1127–40.

17. EQUALITY AND COMMUNITY

Gerald A. Cohen (1941–2009) was a Canadian-born political philosopher. His "camping trip" is in: *Why Not Socialism?* (Princeton: Princeton University Press, 2009). See also his books: *Rescuing Justice and Equality* (Cambridge, MA: Harvard University Press, 2008); *If You're an Egalitarian, How Come You're So Rich?* (Cambridge, MA: Harvard University Press, 2009); *On the Currency of Egalitarian Justice, and Other Essays in Political Philosophy* (Princeton: Princeton University Press, 2011).

"From each according to his ability, to each according to his need": the old socialist formula comes from Marx's *Critique of the Gotha Program* (1875). Margaret Atwood revised the formula in *The Handmaid's Tale* (1985).

Reciprocity is an old concept in ethics. Reciprocity begins with Kant and Hegel and proceeds with T.H. Green to find its place among Canadian idealist philosophers a century ago. Reciprocity, for Green, was "the reciprocal claim of all upon all to be helped in the effort after a perfect life": T.H. Green, *The Prolegomena to Ethics*, ed. A.C. Bradley, 3rd ed. (Oxford: Clarendon Press, 1890), 308; cited in Avital Simhony, "T.H. Green: The Common Good Society," *History of Political Thought* 14, no. 2 (Summer 1993), 229.

John Clark Murray said, "All actions bring men and women into 'manifold reciprocity' with their fellows; it is the 'imperious necessity' of community": *A Handbook of Psychology* (London: Alexander Gardner, 1888), 242.

Robert MacIver, in his 1917 book on community, said, "The 'law of reciprocity' is the 'deepest question of social philosophy'": *Community: A Sociological Study* (London: Macmillan, 1917), 241.

Gerald Cohen said, "Communal reciprocity is the anti-market principle according to which I serve you not because of what I can get in return by doing so but because you need or want my service, and you, for the same reason, serve me": *Why Not Socialism?* (Princeton: Princeton University Press, 2009), 39.

Pierre Rosanvallon discusses homo reciprocans and "reciprocity as equality of involvement" as alternatives to homo economicus and "rational choice theory" in: *The Society of Equals,* trans. Arthur Goldhammer (Cambridge, MA: Harvard University Press, 2013), 269–76.

Reciprocity—the practice of exchanging with others in the community for mutual benefits—is part of Indigenous knowledge in Canada and part of an ancient wisdom upon which all Canadians may draw. Reciprocity occupies a moral and political plane far beyond the heritage of utilitarianism and its intellectual descendants. Recent explorations of the reciprocity concept include: Patrici Calvo, *The Cordial Economy: Ethics, Recognition and Reciprocity* (Cham: Springer, 2018); Antti Kujala and Mirkka Danielsbacka, *Reciprocity in Human Societies: From Ancient Times to the Modern Welfare State* (Cham: Palgrave Macmillan, 2019).

18. SUCKING UP WEALTH

There is a biography of Joseph Flavelle: Michael Bliss, *A Canadian Millionaire: The Life and Business Times of Sir Joseph Flavelle* (Toronto: Macmillan, 1978).

Heterodox economics tries to rebuild economics on unorthodox foundations. There are many books and among the more accessible are: Ha-Joon Chang, *23 Things They Don't Tell You about Capitalism* (New York: Allen Lane, 2010); Kate Raworth, *Doughnut Economics: Seven Ways to Think Like a 21st-Century Economist* (Chelsea Green Publishing, 2017); Andrew Mearman, Sebastian Berger and Danielle Guizzo, *What Is Heterodox Economics?: Conversations with Leading Economists* (Abingdon: Routledge, 2020).

Among the many books on the Occupy movement: Mark Bray, *Translating Anarchy: The Anarchism of Occupy Wall Street* (Winchester: Zero Books, 2013); Nathan Schneider, *Thank You, Anarchy: Notes from the Occupy Apocalypse* (Berkeley: University of California Press, 2013); Sanford F. Schram, *The Return of Ordinary Capitalism: Neoliberalism, Precarity, Occupy* (Oxford: Oxford University Press, 2015).

On climate change and the lessons of war: Seth Klein, *A Good War: Mobilizing Canada for the Climate Emergency* (Toronto: ECW Press, 2020). For an attempt to answer the plumber's question at the end of Chapter 18 see Sager, *Inequality in Canada: The History and Politics of an Idea* (2020).

19. RISING ABOVE MYTHS

Epictetus (50–135 AD) was a Greek Stoic philosopher.

"Free markets foster environmental friendliness": Michael Rieger, "In the Fight against Climate Change, Free Markets Are Our Biggest Ally," *Foundation for Economic Education* (9 November 2018), https://fee.org/articles/in-the-fight-against-climate-change-free-markets-are-our-biggest-ally/.

"The natural differences in ability": Michael Shermer, "Utopia is a dangerous ideal. We should aim for 'protopia,'" *Quartz* (4 April 2018).

"Inequality is the mechanism": Manuel Hinds, "Inequality can be a good thing," *Quartz* (24 June 2013), https://qz.com/96836/inequality-can-be-a-good-thing/.

Among the books from Mariana Mazzucato are: *The Entrepreneurial State: Debunking Public vs. Private Sector Myths* (London: Anthem Press, 2014); *The Value of Everything: Making and Taking in the Global Economy* (New York: Public Affairs, 2018); *Mission Economy: A Moonshot Guide to Changing Capitalism* (London: Allen Lane, 2021).

SOURCES

"They are casting their problems": Margaret Thatcher, in an interview with *Woman's Own* magazine in 1987.

"No man is an island": John Donne, *Meditation XVII*.

"It seems to me, however": Harry G. Frankfurt, *On Inequality* (Princeton: Princeton University Press, 2015), 3.

"A deep and important contribution": Daron Acemoglu, "The Crisis of 2008: Structural Lessons for and from Economics," *Centre for Economic Policy Research, Policy Insight* no. 28 (2009); cited in Jean Tirole, *Economics for the Common Good*, trans. Steven Rendall (Princeton: Princeton University Press, 2017), 48–9.

There is a substantial body of writing on ethics and economics. A critical stance is taken by the economist Jonathan Aldred in: *Licence to Be Bad: How Economics Corrupted Us* (London: Allen Lane, 2019); *The Skeptical Economist: Revealing the Ethics Inside Economics* (Abingdon: Earthscan, 2009).

Another economist's critique of neoliberalism is: Ha-Joon Chang, *23 Things They Don't Tell You about Capitalism* (New York: Allen Lane, 2010).

A fascinating and accessible book by one of the leading philosophers is: Michael J. Sandel, *What Money Can't Buy: The Moral Limits of Markets* (New York: Farrar, Straus and Giroux, 2012).

Mark Kingwell takes on related issues in: *On Risk Or, If You Play, You Pay: The Politics of Risk in a Plague Year* (Windsor: Biblioasis, 2020).

Economics students have been the subject of many attitudinal surveys and experimental studies. Much of the discussion relates to whether they are self-selecting for selfishness or whether studying economics induces a higher degree of selfishness. Among the many publications: Philipp Gerlach, "The games economists play: Why economics students behave more selfishly than other students," *PLoS One* 12, no. 9 (2017); Robert H. Frank, Thomas Gilovich and Dennis

T. Regan, "Does Studying Economics Inhibit Cooperation?" *Journal of Economic Perspectives* 7, no. 2 (Spring 1993), 159–71; Long Wang, Deepak Malhotra and J. Keith Murnighan, "Economics Education and Greed," *Academy of Management Learning & Education* 10, no. 4 (February 2012); Bjorn Frank and Gunther G. Schulze, "Does economics make citizens corrupt?" *Journal of Economic Behavior & Organization* 43, no. 1 (September 2000), 101–13; Yoram Bauman and Elaina Rose, "Why Are Economics Students More Selfish than the Rest?" (IZA Discussion Paper no. 4625, 2009).

Concern over these and other findings encouraged the growth of ethics courses in Economics and Business programs. Ethics courses then prompted research on the impact of these courses on students' attitudes.

"Philanthropy, which translates": Robin Rogers, "Philanthropy and Inequality," *Stanford Social Innovation Review* (Summer 2013).

On charitable giving as a percent of the GDP: "Gross Domestic Philanthropy: An International Analysis of GDP, Tax and Giving," *Charities Aid Foundation* (2016), https://www.cafonline.org/docs/default-source/about-us-policy-and-campaigns/gross-domestic-philanthropy-feb-2016.pdf.

A recent critical gaze on philanthrocapitalism is: Anand Giridharadas, *Winners Take All: The Elite Charade of Changing the World* (New York: Alfred A. Knopf, 2018).

On the inverse relationship between inequality and charitable giving: Nicolas J. Duquette, "Inequality and Philanthropy: High-Income Giving in the United States 1917–2012," *Explorations in Economic History* 70 (October 2018), 25–41. See also: Peter Bloom and Carl Rhodes, *CEO Society: The Corporate Takeover of Everyday Life* (London: Zed Books, 2018).

On the politics of belonging, see, among others: George Monbiot, *Out of the Wreckage: A New Politics for an Age of Crisis* (London: Verso, 2017).

20. ALTRUISM AND A GAME

The dictator game and 20–30 percent: Dan Ariely, "Why We Try So Hard to Escape Our Humanity," *New York Times* (25 August 2018). His book is: *Predictably Irrational: The Hidden Forces That Shape Our Decisions* (New York: HarperCollins, 2008).

The dictator-game experiments became very sophisticated, and today we can even read about the effect of classical music on game outcomes: Aurora García-Gallego et al., "The Heaven Dictator Game: Costless taking or giving," *Journal of Behavioral and Experimental Economics* 82 (October 2019).

The idea that human nature entails a strong barrier to egalitarian reforms and even to the transformative powers of reason has a long and distinguished lineage. For a summary, see: Paula Casal, "Distributive Justice and Human Nature," in *The Oxford Handbook of Distributive Justice* (2018), 259–82.

Is there a genetic component to empathy?: Martin Melchers et al., "How heritable is empathy? Differential effects of measurement and subcomponents," *Motivation and Emotion* 40 (August 2016), 720–30; Varun Warrier et al., "Genome-wide analyses of self-reported empathy: correlations with autism, schizophrenia, and anorexia nervosa," *Translational Psychiatry* 8 (March 2018); James W. H. Sonne and Don M. Gash, "Psychopathy to Altruism: Neurobiology of the Selfish–Selfless Spectrum," *Frontiers in Psychology* 9 (April 2018).

On homo caritatis: the not-uncontroversial feminist ethics of care has strong advocates. See, for instance: Virginia Held, "The Ethics of Care," in *The Oxford Handbook of Distributive Justice* (2018), 213–34; Joan C. Tronto, *Caring Democracy: Markets, Equality, and Justice* (New York: NYU Press, 2013).

Among the many works on the feminist confrontation with economics: Martha A. Fineman and Terence Dougherty, eds., *Feminism*

Confronts Homo Economicus: Gender, Law, and Society (Ithaca: Cornell University Press, 2005).

Homo caritas, of course, has a central place in Catholic theology.

Inequality aversion: Erte Xiao and Cristina Bicchieri, "When Equality Trumps Reciprocity," *Journal of Economic Psychology* 31, no. 3 (June 2010), 456–70; C. T. Dawes et al., "Egalitarian motives in humans," *Nature* 446 (April 2007), 794–6; Alex Shaw and Kristina R. Olson, "Children Discard a Resource to Avoid Inequity," *Journal of Experimental Psychology* 141, no. 2 (2012), 382–95; Ernst Fehr, H. Bernhard and B. Rockenbach, "Egalitarianism in Young Children," *Nature* 454 (2008), 1079–83.

Americans on the ideal ratio of CEO pay: Sorapop Kiatpongsan and Michael I. Norton, "How Much (More) Should CEOs Make? A Universal Desire for More Equal Pay," *Perspectives on Psychological Science* 9, no. 6 (November 2014), 587–93.

Attitudes on wealth inequality in Canada: "The Wealth Gap: Perceptions and Misconceptions in Canada," *Broadbent Institute* (2014); "Progress Summit 2017," *Broadbent Institute* (2017); summarized in: Ethan Cox, "Poll: Canadians deeply concerned by growing inequality," *Ricochet* (7 April 2017).

For a comparative analysis of attitudes toward income distribution: Lars Osberg and Timothy Smeeding, "'Fair' Inequality? Attitudes toward Pay Differentials: The United States in Comparative Perspective," *American Sociological Review* 71, no. 3 (2006), 450–73.

Inequality is not accepted: even in what may seem the most unlikely of places, a redistributive program can win widespread support. An example is the 2016 campaign of Bernie Sanders in the United States: George Monbiot, *Out of the Wreckage: A New Politics for an Age of Crisis* (London: Verso, 2017), 166–74; Becky Bond and Zack Exley, *Rules for Revolutionaries: How Big Organizing Can Change Everything* (White River Junction: Chelsea Green Publishing, 2016).

21. AN EQUALITY MANIFESTO

Equality of consideration of interests: a concept developed by the philosopher Peter Singer, especially in the context of animal rights and "speciesism."

On climate change and distributive justice, there is a valuable summary of the issues in: Simon Caney, "Climate Change," in *The Oxford Handbook of Distributive Justice* (Oxford: Oxford University Press, 2018), 664–88.

On emissions permits: Olle Torpman, "The Case for Emissions Egalitarianism," *Ethical Theory and Moral Practice* 22 (2019), 749–62.

Permit trading, of course, has generated research and controversy.

On tax evasion, see, for instance: Emmanuel Saez and Gabriel Zucman, *The Triumph of Injustice: How the Rich Dodge Taxes and How to Make Them Pay* (New York: W.W. Norton, 2019); Alain Deneault, *Canada: A New Tax Haven* (Vancouver: Talonbooks, 2015).

"If you do not change direction": a saying attributed to Gautama Buddha.

Egalitarian justice has experienced a revival in recent years, from various theoretical and political positions. Among others: Erik Olin Wright, *Envisioning Real Utopias* (London: Verso, 2010); Iwao Hirose, *Egalitarianism* (Abingdon: Routledge, 2015); Pierre Rosanvallon, *The Society of Equals* (Cambridge, MA: Harvard University Press, 2013); Jonathan Rothwell, *A Republic of Equals: A Manifesto for a Just Society* (Princeton: Princeton University Press, 2019).

David Harvey makes a case for radical egalitarianism in: *The Enigma of Capital: And the Crises of Capitalism* (Oxford: Oxford University Press, 2010).

Socialist alternatives are experiencing a revival, at both scholarly and popular levels. For instance: Axel Honneth, *The Idea of Socialism: Towards a Renewal* (Cambridge: Polity Press, 2017); Bhaskar Sunkara, *The Socialist Manifesto: The Case for Radical Politics in an Era of Extreme Inequality* (New York: Basic Books, 2019).

The microcredit movement (controversially) presents itself as an egalitarian alternative to failed capitalism and to inequality. For instance: Muhammad Yunus, *A World of Three Zeros: The New Economics of Zero Poverty, Zero Unemployment, and Zero Net Carbon Emissions* (New York: Public Affairs, 2017).

THANKS AND FINAL THOUGHTS

I DIDN'T WRITE this book myself. I wrote a draft during the first three months of 2020. After that, this strange little thing began to evolve as it responded to a new environment—a gathering of friends, family, and critics at arm's length, each of them unknown to most of the others. This book is the work of a large community, and to all in that community, I express sincere gratitude: Fran Baskerville, Peter Baskerville, Gregory Blue, Cheryl Coull, Leo Eutsler, Deborah Gesensway, Corina Lee, Phil Lew, John Lutz, Cathie Monahan, Pat Monahan, Grant Morrison, Nalini Elisa Ramlakhan, Veronica Strong-Boag, David Wilson, Zoë Sager, Jean Anne Wightman, and an anonymous reader from FriesenPress. Angus Morrison offered valuable information about plumbing. Although the members of this informal writing workshop did not gather in one place, each spoke about the book through me, and each had a part in its evolution. Where the end result fails to respond adequately to the many comments and challenges from these readers, the responsibility is mine.

Jean Anne—my constant and loving luminary—urged the value of illustrations, and she was right, as she so often is. But neither of us anticipated the transformation that Hanna Melin would effect. Hanna read the book, understood what R (the professor) and A (the plumber) were saying, and bestowed upon their conversation her penetrating and empathetic gaze. Her contribution lies not in parts of the book, but in the whole.

The Professor and the Plumber is the tiny offspring of a distended monster: *Inequality in Canada: The History and Politics of an Idea*

(Montreal: McGill-Queen's University Press, 2020). Hence, my oft-repeated self-mockery: I conclude my career by writing one book that is publishable but unreadable, and another book that is readable but unpublishable. This jest may be unfair to both books, of course. A number of academics will attend to (and probably carve up) the monster; and the little offspring is now alive in print. The truth behind this jest is that thirteen publishers (and one agent) declined to consider this book. To be precise, they did not reject the book, because they did not even look at it. They received my detailed proposal and lost interest at that stage (many did not bother to reply). Why the lack of interest? There is certainly more than one reason, but a theme emerges. This book is not about topics currently in fashion. It is not about the COVID pandemic, or racism, or the politics of sexual identity. One publisher's representative stated bluntly: there are already too many books about economic inequality.

That missed opportunity is their loss, not mine. I have had the reward of publishing with the guidance of the skilled professionals at FriesenPress. I have been given an escape from the glacial pace of publication in Canada. And I have gained the experience of guiding this book into print and into the book market. For this experience and these rewards, I am truly grateful.

The Professor and the Plumber was born out of a personal history that will be apparent to those who know me well. I was raised in a family of siblings and extended kin to whom I have always been close. Here, in this large family, I learned about community, equality, and values of the common good. I am the only university professor in my extended family. Often my relatives have asked about my work. They know that I am not only a teacher of adults, but also a researcher and writer. Some have inquired further about what I do. This book is my answer, for those who may be interested: here is what I do. Here is what I think about. This book is for my sisters, my children, and grandchildren, and for my cousins. I hope you do not find it too off-putting, even if much of it is challenging. Some of it is difficult, and it is difficult for me too. It must be so. Nothing of value in life comes easily.

THANKS AND FINAL THOUGHTS

This book is not written for scholars, although I hope that many will read it. They will find much to quarrel with here. Philosophers, especially, are likely to squirm over and complain about my radical reductions and simplifications. To such complaints, I have one answer: where have you been? Why have you left the wider dissemination of important ideas in distributive justice to a historian? A few philosophers have written for wider audiences. But Michael J. Sandel and the Canadian Mark Kingwell are exceptions. Apart from books by Sandel and Kingwell, there has been virtually no attempt to draw egalitarian or communitarian principles of distributive justice into a politics that is useable by politically progressive non-academic readers. Even the lucid writing of Gerald Cohen veers toward complex argumentation that many non-academic readers will find impenetrable. And when economic inequality appears in our media, as it so often does, who gets to speak? Economists have pushed philosophers off the stage. Who today would know that the study of inequality has always been, as it remains, the work of philosophers?

I tell my story through dialogue for more than one reason. Dialogue is the mode in which we exchange and sharpen our ideas. It is the mode in which complex ideas may appear in short "sound bites." It is the mode in which we debate ideas in university seminars; and it is the mode in which cousins interact and sometimes debate serious issues with each other at family reunions. Dialogue is commonplace in fiction: we know the characters through their speech and their internal self-talk. Plato may be the greatest, but he is not the only writer to use dialogue to propel thought and action in non-fiction. I have been inspired by Jostein Gaarder (*Sophie's World: A Novel About the History of Philosophy*) and other modern practitioners. I remain an unskilled novice in this art. I hope, however, that when R and A speak, their ideas take wing and make sense.

The conversations in this book are intended to challenge and provoke, and they will do so in many ways. Readers meet two people, and questions arise. Who are these people? What is their relationship to each other? Answers to the "mystery" are slow to appear; one must read on. In the gaps, the puzzled reader may insert assumptions or guesses (already, prior to publication, the opening pages of this book have triggered an astonishing range of responses). Professor and plumber: what do these connote? At the beginning, we learn more about the plumber, who is indeed male; the professor is reticent. The professor proposes the topic; the plumber lays down conditions for the conversation; the professor accepts his conditions. His hangover, if there is one at all, does nothing to blunt his acuity; he makes substantive observations about inequality. He asks the most important question of all (what is the purpose of a discussion about inequality?); and he challenges his scholar-cousin with an acerbic reproach (sometimes you just talk to yourself). At the end of Chapter 18 he asks a profoundly challenging question about the relevance of history. Does one of these two speakers think more deeply than the other? If so, which one? One drinks wine, the other beer (but not always): what, if anything, do you infer from that fact? One reads many books, the other reads a few: do we infer anything at all about the extent of knowledge possessed? Let us hope that we infer nothing, even unconsciously, about capacities for moral reasoning. What assumptions and values do we bring from our own positions, whether of class, or gender, or ethnicity, or any other part of our identity?

When we eavesdrop on these conversations, it is always clear who is speaking. Until a certain point in the final chapter. Thereafter, in the reciprocity of their egalitarian space, two voices speak as one. Only then do we hear their names.

This book is about that egalitarian space. If there is to be a collective transcendence of inequality and its associated evils—if we are to find an escape from the suffocations of neoliberalism—progressive forces must find a new language of equality and a new politics of community. And professors should do a better job of conversing with plumbers and also with those at lower levels of the income distribution. In doing so, we should begin with ethics and principles of justice. This book is, I hope, a contribution to that great and necessary project.

ABOUT THE AUTHOR

ERIC SAGER has taught history for nearly 50 years. His books include *Inequality in Canada: The History and Politics of an Idea* (2020), *Seafaring Labour* (1989), and *Ships and Memories* (1993). He lives in Victoria, British Columbia.

HANNA MELIN fell in love with drawing at an early age, a passion that took her to London's Royal Academy of Art where she earned a master's degree in visual communication. Born in Sweden, Melin lives in East London, where she works as an illustrator and designer.

CPSIA information can be obtained
at www.ICGtesting.com
Printed in the USA
BVHW020815150921
616188BV00012B/37